T0332173

COMMUNITIES OF INNOVATION

How Organizations Harness Collective Creativity and Build Resilience

COMMUNITIES OF INNOVATION

How Organizations Harness Collective Creativity and Build Resilience

Editors

Patrick Cohendet
HEC Montréal, Canada

Madanmohan Rao
YourStory Media, India

Émilie Ruiz
Strasbourg University, France & Lorraine University, France

Benoit Sarazin
Innovation Consultant, France

Laurent Simon
HEC Montréal, Canada

NEW JERSEY • LONDON • SINGAPORE • BEIJING • SHANGHAI • HONG KONG • TAIPEI • CHENNAI • TOKYO

Published by

World Scientific Publishing Co. Pte. Ltd.
5 Toh Tuck Link, Singapore 596224
USA office: 27 Warren Street, Suite 401-402, Hackensack, NJ 07601
UK office: 57 Shelton Street, Covent Garden, London WC2H 9HE

Library of Congress Cataloging-in-Publication Data
Names: Cohendet, Patrick, editor. | Rao, Madanmohan, editor. | Émilie, Ruiz, editor.
Title: Communities of innovation : how organizations harness collective creativity and
build resilience / editors, Patrick Cohendet, HEC Montréal, Canada,
Madanmohan Rao, YourStory Media, India,
Ruiz Émilie, Strasbourg University, France & Lorraine University, France,
Benoit Sarazin, Innovation Consultant, France, Laurent Simon, HEC Montréal, Canada.
Description: USA : World Scientific, 2021. | Includes bibliographical references and index.
Identifiers: LCCN 2021001179 | ISBN 9789811234279 (hardcover) |
ISBN 9789811234286 (ebook) | ISBN 9789811234293 (ebook other)
Subjects: LCSH: Creative ability in business. | Communities of practice. |
Business enterprises--Technological innovations.
Classification: LCC HD53 .C6536 2021 | DDC 658.4/06--dc23
LC record available at https://lccn.loc.gov/2021001179

British Library Cataloguing-in-Publication Data
A catalogue record for this book is available from the British Library.

Copyright © 2021 by World Scientific Publishing Co. Pte. Ltd.

All rights reserved. This book, or parts thereof, may not be reproduced in any form or by any means, electronic or mechanical, including photocopying, recording or any information storage and retrieval system now known or to be invented, without written permission from the publisher.

For photocopying of material in this volume, please pay a copying fee through the Copyright Clearance Center, Inc., 222 Rosewood Drive, Danvers, MA 01923, USA. In this case permission to photocopy is not required from the publisher.

For any available supplementary material, please visit
https://www.worldscientific.com/worldscibooks/10.1142/12208#t=suppl

Desk Editors: Aanand Jayaraman/Sandhya Venkatesh

Typeset by Stallion Press
Email: enquiries@stallionpress.com

Printed in Singapore

© 2021 World Scientific Publishing Company
https://doi.org/10.1142/9789811234286_fmatter

About the Editors

Patrick Cohendet is Professor at HEC Montréal in the Department of international business. He is also Co-Director of Mosaic, a creativity and innovation hub at HEC Montréal, and Co-editor of *Management International Journal*. His teaching, research and publications focus on the economics and management of innovation, knowledge and creativity. He is the author of more than 80 articles published in peer-reviewed journals and 15 books including Architectures of Knowledge written with Ash Amin in 2004. In 2017, he co-edited *The Elgar Companion to Innovation and Knowledge Creation* published by Edward Elgar Publishing. His research has been published in journals such as *Research Policy*, *Organization Science* and *Industrial and Corporate Change*.

Madanmohan Rao is Research Director at YourStory Media in Bangalore, India. He is also a Charter Member at entrepreneurship support organization TiE Bangalore, and Knowledge Summit Committee Member at the Confederation of Indian Industry (CII). Madan is the editor of five book series spanning 15 titles, including *The Knowledge Management Chronicles, The Asia-Pacific Internet Handbook* and *World of Proverbs.* He has given talks and lectures in 90 countries around the world. His hobbies include world music, jazz and craft beer appreciation. Madan graduated from IIT Bombay (BTech, Computer Science) and University of Massachusetts at Amherst (PhD, Communications).

Émilie Ruiz is Associate Professor at Strasbourg University, teaching at IUT Louis Pasteur and member of the Creativity — Science — Innovation department at BETA research laboratory. Her research deals with open innovation and crowdsourcing, with a particular focus on the adoption process of crowdsourcing for innovation strategies, and on the cultural and creative industries. Her research is published in *R&D Management*, *Management International,* etc.

Benoit Sarazin is a consultant and speaker specializing in disruptive innovation. Originally an expert in the IT and Telecoms sector at Hewlett-Packard, in the heart of Silicon Valley, where he was Marketing Director, he has applied his experience of managing innovation to advise innovative companies in multiple sectors. At the head of Farwind Consulting since 2001, he is the author of the blog "Innovation de rupture" (www.benoitsarazin.com), and of the book *Innovation de rupture: guide pour disrupter votre marché* (EMS, 2019). In the course of his missions with executive committees, Benoit has developed an approach that capitalizes on the contribution of innovation communities.

Laurent Simon is Professor at HEC Montréal in the Department of Entrepreneurship and Innovation. He is also Co-Director of Mosaic, a creativity and innovation hub at HEC Montréal. He teaches courses on the management of innovation and creativity, design thinking and business models. His research focuses on the organization, management and performance of creative and innovative processes at the individual, collective, organizational and territorial levels. In 2017, he co-edited *The Elgar Companion to Innovation and Knowledge Creation* published by Edward Elgar Publishing. His research has been published in journals such as *Organization Science*, *Journal of Economic Geography*, *Industry and Innovation* and *Management International*.

© 2021 World Scientific Publishing Company
https://doi.org/10.1142/9789811234286_fmatter

About the Contributors

Lusine Arzumanyan is a consultant in strategy and business development. She holds a doctoral degree in Management from Jean Moulin Lyon 3 University. Her doctoral dissertation analyzes the implementation of communities of practice in the field of innovation in multinational corporations. The empirical study of her doctoral thesis is based on the case of the French Groupe SEB. She is an associate researcher at ESDES The Business School of UCLY. Her research and teaching interests focus on areas of innovation management and strategy.

Hugues Boulenger is a disruptive expert of talent identification, and a business innovation mentor. He and his team developed a disruptive way to find people's unique talents with incredible precision. Founder of the consulting agency "Réussir dans le plaisir", he and his team help people find their creative genius and organizations collaborate focusing on the genius of each team member. They also provide consultancy services on innovation, business strategy and digital Marketing strategies. After Two Masters in Management and Strategy and Organizations from EDHEC Business School Grande Ecole, and a degree in psychology, Hugues worked in strategy and customer profitability optimization strategies in big agencies and then in consulting in innovation with Benoit Sarazin, before he started Réussir dans le Plaisir. He is passionate about Talent, Innovation, Business and Humor.

Sébastien Brion is Professor of Management in Logistics and Operation research laboratory in Aix Marseille University. He is associate research fellow at the IREGE research laboratory, which he led for many years. His research interests focus on digital innovation, organizational forms conducive to innovation, especially in open innovation context. His work has been published in many ranked journals including *Journal of Small Business Management*, *R&D Management*, *International Journal of Project Management*, *M@n@gement* and *International Journal of Innovation Management*.

Tristan Cenier is a PhD in Neuroscience from University Claude Bernard in Lyon, France. His research investigated the biological support for odor information encoding and representation in the brain. It was conducted partly in France and in the USA.

After realizing a researcher's mindset and work methods were particularly fit to manage industrial innovative projects, he got a Master's degree in "Innovative Project Engineering" and joined furniture manufacturer Schmidt Groupe to conduct long-term product development, investigate new fields of activity deployment and organize collective intelligence and creativity within the company.

Jean-Yves Couput is Director of innovation at the Footwear and Business division of Amer Sports Corporation. He is also the director of the ME:sh project, an internal startup within the Salomon brand, and the pilot of the brand's relationship with communities. He has been Professor of Physical Education and a graduate of the Institute of Political Studies in Toulouse. After having been a professional cyclist in the late 1980s, Jean-Yves Couput turned to a career as a sports journalist and then as a Community Marketing expert with the Salomon brand in Annecy, France. Around five years in the United States, followed by four years in Finland, allowed him to theorize and implement various models of community dynamics in the world of sports.

Florence Crespin-Mazet is an Affiliate Professor of Business-to-Business Marketing at Kedge Business School (KBS), France, and has specialized in business-to-business marketing, solutions selling and emergent collaborative forms of innovation (communities, networks). She is an active member of the KCO Community (Knowledge Communities Observatory) from KBS. She carried out her PhD at the University of Manchester on the issue of co-development as a marketing strategy in project business. Florence also holds an MBA from Illinois State

University (USA), as well as an MSc degree from ESSCA business school. Her main research interests include the evolution from transactional to relational approaches and customercentric organization in project businesses as well as the role of communities and networks in innovation & business development. Florence has specific expertise in the construction industry. Florence has worked with various industrial and service organisations including Air Liquide, Arcadis, Bouygues, EDF, Kéolis, Laerdal Medical, Saint-Gobain, Spie batignolles, Safran, Sartorius Stedim, Schneider-Electric.

Odile de Saint Julien has a PhD in Management Sciences. A Senior Strategy Consultant and Specialist in Innovation Ecosystems in Life Sciences, she has participated in the development and implementation of collaborative strategies for scientific research. In South Africa, she co-created and directed an Experimental Research Laboratory. In Mozambique, she created the "May Workshops" to accompany, women "creators" of micro-businesses in order to promote local handicrafts. In India, she created an incubator to promote the creation of a 30th startup in the field of eye surgery. Since 2016, Odile has been an assistant professor at KEDGE. She teaches strategy, open innovation and entrepreneurship. She participates, as an expert, in different commissions for the evaluation of entrepreneurial projects (COPIL PEPITE Paca-Est).

Coline Delmas was the Community Program Manager of Schneider Electric for three years. She developed the internal communities by supporting the creation and animation of 170 communities of practice and leading the network of 200 community leaders. She helped improve the usages of the collaboration tools and coached the community leaders to improve their practice. The quest for innovative facilitation practices is at the heart of this program. She is currently an active member of the core team of the Specifiers Community.

Karl-Emanuel Dionne is an Assistant professor at HEC Montréal in the department of entrepreneurship and innovation. His work focuses on topics of new forms of organizing for innovation, interdisciplinary and interorganizational collaboration, open innovation events and ecosystem emergence at the intersection of different sectors. He is also interested in the role innovators play in the transformation of existing patterns of behaviors to develop novel ideas that span established organizational and disciplinary boundaries. His research particularly takes place in the context of the digital transformation of healthcare. His work has been published in outlets such as *Organization Studies* and *Research in the Sociology of Organizations*.

Claudia Folco has been Community Program Manager for Schneider Electric since 2018, replacing Coline Delmas. She supports 220 communities along their full lifecycle, animating at the same time the Network of 250 community leaders. She ensures best practices sharing around community management and collaborative tools usages, providing also training and on-demand support. She is convinced of the strong value that communities bring to innovation.

Romain Gandia is Associate Professor at University of Savoie Mont Blanc (France) and permanent researcher in the IREGE laboratory. He holds a PhD in Management Sciences with a specialization in innovation management. His current research interests mainly focus on strategic and organizational behaviors of innovative companies in creative and digital industries. He has published in international academic journals such as *Journal of Small Business Management*, *Strategic Change*, *Journal of Business Strategy*, *European Business Review*, and *Creativity and Innovation Management*.

Karine Goglio-Primard is an Associate Professor of BtoB Marketing at Kedge Business School, Toulon, France. She received her PhD in economics from the University of Nice-Sophia Antipolis. Karine also holds a Master's from the university of Paris 1 Panthéon Sorbonne. Her research focuses on communities of practice that emerge in organizations and are cultivated to nurture innovation and business development. She founded the Knowledge Communities Observatory (KCO) at Kedge Business School, which brings together companies (Crouzet, ENGIE, Expleo, Laerdal Medical, Schneider Electric, Sartorius Stedim, Spie Batignolles, Amallte and Saint-Quentinois Agglomeration community) and expert researchers on communities. Within the KCO, she analyzes how the formal structures of companies integrate the production of communities into their innovation processes.

Erik Grab is Vice President of Michelin in charge of Strategic Anticipation and Co-Innovation for the Michelin Group since 2012. He was previously Global Vice President Marketing for the two main

Business Units of Michelin: the BtoB one and the BtoC one. He also created in 2014 the Movin'On LAB, a strategic anticipation and co-innovation ecosystem gathering over 300 international private corporations and public entities as well as academics and international organizations, based in Paris and now also in Montreal.

Before joining Michelin in 2003, Erik had founded a Business Consulting Agency ICON which he subsequently sold to WPP, world leader in advertising and marketing services (300 employees when it was sold). Erik began his career in sales, then turned to marketing consultancy and market studies, including within the German Institute GfK before becoming the company's French branch CEO.

Erik is a strategic anticipation and innovation unconditional enthusiast and regularly speaks on university campuses and international forums on the future of mobility and more generally on innovation processes and methodologies. He is also a member of the Futuribles International Board, President of the third research initiative "Economy & Climate" at the University of Dauphine and President of the Strategic Orientation Committee of ICN. He is also the new Mobilities Project Coordinator for the PFA (Plateforme de la Filière Automobile Française) and Vice President of the "Mobility Factory" created with ADEME in France, Canada, and soon in Africa. He is one of the four qualified personalities in France who are advising the Defense Innovation Agency and is a member of the innovation steering committee of the French Army.

Louis-Pierre Guillaume is a consultant and lecturer, specialized in Digital Transformation & Knowledge Management. At the head of Amallte (https://www.amallte.com), he helps organizations to foster their Digital Transformation and Collaborative Innovation to enhance the value of knowledge. He has 25 years of practical experience in knowledge management in big multinational companies (Schlumberger, Areva, Schneider Electric...). He was the Director of the Schneider Electric Knowledge Management Office, where he led the governance and global implementation of knowledge sharing, in particular the 200+ communities of practice and the internal Wikipedia. He is the Vice President of CoP-1 (www.cop-1.net), the association of KM practitioners in France. He teaches at KEDGE BS, Sorbonne Nouvelle University and CNAM INTD, and gives lectures in professional and academic environments. He has to his credit 90+ presentations, publications and contributions. (http://conferences.amallte.com)

Claude Guittard is an Associate Professor at the BETA laboratory (Research centre on theoretical and applied economics) of the University of Strasbourg, France, and studies the dynamics of knowledge creation and adoption in various types of settings. Expert in community management with a special focus on crowdsourcing in the digital area, he investigates the interplay between isolated creative individuals (experts), on the one hand, and large mobilizations of crowds towards a common objective (innovation), on the other.

Dominique Levent is Creativity Vice President of Renault Nissan Mitsubishi Alliance and in charge of the Renault Institute for Sustainable Mobility. She is also an Expert Leader, "Innovation Patterns". An Arts & Métiers Engineer, she was successively in charge of research on comfort, the first woman Product Manager for Clio, Product Prospective Planning Manager where she imagined the first Scenic and the first Kangoo, Head of Forward Planning for Commercial Services, to finally join the Innovation Department in 2007 where she is now developing projects to boost the Alliance's Innovation capacities.

Patrick Llerena has been a Professor in Economics since 1988 at the Faculty of Economics and Management, University of Strasbourg and a researcher at Bureau d'Economie Théorique et Appliquée (CNRS UMR n°7522), University of Strasbourg, France. He was CEO of the Foundation University of Strasbourg from 2009 to 2015, and former vice-director and director of the BETA from 1991 to 2008. He also has long-term experience in university management, such as vice-president of the University Louis Pasteur, Strasbourg, in charge of finance, and then of industrial relations. Since March 2017, he has been charged by the University President to develop "Student entrepreneurship" as a transdisciplinary track of study at the university level. He has published numerous articles in academic journals and edited books in the following fields: Economics and Management of Creativity and Innovation; Economics of Science; Theories of the firm and of organization; Scientific and Innovation Policies and Decision theory under uncertainty. In particular two books have ben published: *Technology policies in a Knowledge-based Economy: theories and practices*, with Mireille MATT (eds), Springer Verlarg, 2005, and *Option valuation for energy issues*, with Katrin OSTERTAG, Alban RICHARD (eds), FhG Verlag, Stuttgart, 2004. He was, from 2005 to 2011, co-coordinator of a Network of Excellence funded by the EU: DIME

"Dynamics of Institutions and Markets in Europe". He led a set of large European and national projects funded by the ESF, the ANR, etc… He has been member of the "Expertkommission für Forschung und Innovationen" (EFI), Berlin, Germany, from 2009 to 2013. He was the first member of the Higher Level Economic Policy Expert Group "Innovation for Growth" (I4G) until 2013, then of the "Research, Innovation, and Science Policy Experts High Level Group" (RISE), Brussels, until 2017. He is heading a Chair in Management of Creativity at the University of Strasbourg, supported by the Fondation Université de Strasbourg and a group of local companies. Since mid-1980s, he has published in prestigious internationals journal in his fields of competences.

Zoé Masson is a CIFRE doctoral student at CERAG, a research laboratory of the University of Grenoble Alpes, and works at the Ixiade research institute in Grenoble. After a Master's degree on the dynamics and use of new media and digital, she is working on a thesis on the identification of methods and tools to ensure and develop creativity and dynamism within online communities.

Guy Parmentier is Professor at the Grenoble Alpes University and a researcher at CERAG. He has managed and participated in European and national research projects on online creativity and the firms' facing virtual environments. Today he is developing research on business models of creative and digital industries and on the development of organizational creative capacities. He also managed the video game production studio Galilea in Grenoble and Montreal for 8 years.

Luc Sirois holds a Bachelor's degree in electrical engineering from McGill University and an MBA from Harvard University. Recognized in Canada and around the world for his creative approach to innovation, M. Sirois is a leader and entrepreneur in digital technology, with investments in numerous startups and non-profit organizations focused on youth, health, science and education. He co-founded the health innovation movement Hacking Health as well as its digital health accelerator and pre-seed fund. He is co-founder of Resonant Medical, now Elekta Canada, a leading manufacturer in the field of radiation oncology and image-guided treatments. He has also served as Vice President of Consumer Health at TELUS Health, Telesystem and

Nightingale, and as Manager at McKinsey & Company with offices in Montreal, Toronto, Zurich and Paris. Today, he is Managing Director of PROMPT, a not-for-profit organization that facilitates R&D partnerships between the industry and research institutions to improve the competitiveness of companies in the information and communications technology (ICT), artificial intelligence and other digital technology markets. Mr. Sirois is also strategic advisor to the Minister of the Economy and Innovation of the Quebec government. As such, he currently works on deploying new tech transfer models, on the culture of innovation in institutions, on issues of business creation and scientific entrepreneurship, as well as on the transfer of social innovations and their adoption in society.

Frédéric Touvard has a technical backgrounds in the aerospace field, having worked for 15 years with the AIR LIQUIDE Group. He worked on a variety of projects, including design and installation in the industry as well as work oriented towards research and innovation. He was in charge of coordinating and starting up the oxygen and nitrogen production plant on board the Charles de Gaulle Aircraft Carrier. He was also leading the fuel cells and hydrogen project of Jean-Louis Etienne's North Pole expedition in 2002.

He held the function of COO at Axane developing hydrogen fuel cells technology on new markets, with a team of 40 people for 10 years. He is the Founder and CEO of CENTAURY France & Canada, consulting and coaching Company specialized in disruptive innovation. As a certified coach, for the past 10 years, he has been supporting companies in their transformation, encouraging human Management approaches operational methodologies and tools. Frederic was the Co-leader of the Renault innovation community. He is an Associate Professor in Innovation Management at CNAM Paris and Expert at APM (Association pour le Progrès du Management). Frederic is also the author of *Le manager explorateur*, *le management de projet par enjeux*, un catalyseur d'innovation, Edition Presse Internationale Polytechnique, and the co-author of the section about Renault in *Les communautés d'Innovation*, Edition EMS 2017.

Catherine Thiesse is Operational Director of the KCO (Knowledge Community Observatory). She was previously Director of Knowledge Management and Communities at Schneider Electric. She led the

Competency Strategy for industry teams. Overall, for more than 25 years, she has been leading multi-functions, multi-cultural communities in an international environment, especially within the industrial and purchasing functions at Schneider Electric.

Charlotte Wieder holds a doctoral degree in Industrial Engineering. She is a specialist in innovation, business transformation and international community of practice engagement. She led the digital transformation and sub-community of practices of the Innovation Community of Groupe SEB in France and the growth of the international PlayFutures community on learning through play at the LEGO Foundation in Denmark. More recently, with the community educatefor.life, she launched WOM'UP: an Edtech passion project, aiming at empowering women in rural villages in Morocco, by leveraging technology and local heritage, for a more connected prosperous global village. Since 2019, she has been working as Beyonder at OCP SA in Morocco (mining and chemical industry), supporting the success of strategic and exponential growth initiatives.

© 2021 World Scientific Publishing Company
https://doi.org/10.1142/9789811234286_fmatter

Contents

© 2021 World Scientific Publishing Company
https://doi.org/10.1142/9789811234286_fmatter

Overview

Patrick Cohendet, Madanmohan Rao,
Émilie Ruiz, Benoit Sarazin and Laurent Simon

Every collective book has a history of specific relationships between different contributors. This book on communities of innovation is in fact the result of a formation of a community of academics and practitioners passionate about the growing role and functioning of communities in society. Initiated by Mosaic, a multidisciplinary platform for research and training in the field of creativity and innovation management at HEC Montréal Business School, the network of authors progressively grew through multiple events such as schools of creativity and innovation organized by the Mosaic team in places such as Montréal, Barcelona, Strasbourg, Bangkok, Tokyo, Lille or Grenoble. For instance, Laurent Simon and Patrick Cohendet, co-directors of Mosaic, first met Madanmohan Rao in the creative school of Bangkok in 2015, and since then Madan participated in all the summer schools organized in Montréal and Barcelona. Benoit Sarazin as well as Émilie Ruiz also participated in such events.

Since such programs were unique occasions to share and compare experiences in innovation and creativity management with managers, researchers and designers, it emerged that one of the main results was to highlight the key and growing role played by communities in the formation of innovation. A first book was published in French on this subject, in 2017, coordinated by Benoit Sarazin, Patrick Cohendet and

Laurent Simon, at the Éditions EMS, under the title: *Les communautés d'innovation: de la liberté créatrice à l'innovation organisée*. Out of the 15 chapters that constitute the present book, 11 were presented in the 2017 book. However, they have all been rewritten and revised for this volume in English, taking into account all recent evolution and modifications since then. In particular, the emergence of the pandemic and more generally the role of communities in situations of crises were a strong incentive to add significant new contributions in this domain, such as the chapters by Madanmohan Rao and by Émilie Ruiz *et al.* This book describes the important role played by communities in innovation processes and how organizations can benefit from them. A community brings together individuals who share a common passion for a given area of knowledge and can contribute to innovation at different levels: capitalization of good practices, problem solving, sharing of expertise or development of new and creative ideas. As practitioners and academics increasingly emphasized the needs of collaborative approaches in innovation, they progressively challenged the traditional idea that innovation is mainly generated by hierarchical corporate departments and highlighted the active role that communities play in innovation processes. Regular interaction and best practices sharing among community members constantly generates a wealth of new ideas. This creativity can naturally be considered as a powerful source of innovation for organizations, which brings positive outcomes such as costs and risks' reduction, formation of a reservoir of useful knowledge, new commercial potentials, etc.

The literature has progressively identified many variants of communities such as communities of practice, epistemic communities, communities of interest, virtual communities, etc. These forms of communities differ regarding the type of the specialized activities of knowledge on which they focus. For instance, while communities of practice are centred on the circulation of best practices in a given domain of knowledge (exploitation), epistemic communities are more concerned with the production of new knowledge (exploration). For a given organization, the literature also distinguishes internal communities (employees of a given organization) from external communities (communities of customers, partners, etc.). However, even if managers rub shoulders with some of these enthusiasts every day, the corporate innovation processes do not guarantee them access to the creative pool generated by communities. Collaboration with communities in order to harness their creative potential requires the implementation of new management methods, which are described in this book. One of the main problems for organizations is that community boundaries are much more permeable than firm boundaries: the communities can expand through multiple corporate boundaries, and in the case of virtual communities, the notion of boundary even does not exist. Collaboration with communities requires the implementation of new management methods, which are described in this book.

The role of communities in innovation process is particularly important in the context of crises, as the recent context of pandemic due to COVID-19. In such circumstances, the reaction from formal and rigid structures is generally slow, difficult, inadequate or inefficient. The potential reaction of communities based on improvization, on societal motivation, and on sharing mutual aid is a key factor for resilience of an organization or a society exposed to hazards to resist, absorb, accommodate to and recover from the effects of a hazard in a timely and efficient manner. Thus, communities can be viewed as unique coordination modes for resilience and creativity when facing crises. Multiple examples detailed in this book illustrate how new and creative solutions emerged from communities in the context of the pandemic, and how these solutions paved the way to build new formal structures better adapted to the new economic and social environment.

This book is organized along the following sections. In a first part we present as an introduction a synthesis on the notion of communities and explain their role in the dynamics of innovation. The second part is focused on the analysis of communities of innovation in specific organizations (Ubisoft, Salomon, Schneider Electric, SEB, Schmidt Group). In the third part, we examined the role of communities in industrial ecosystems of multiple organizations (Open Lab Michelin, Renault community). The fourth part focuses on the role of the middleground, as an intermediary platform for mobilizing different types of communities (*Trackmania*, Hacking Health, Afrikaner). Part five specifically examines the role of communities as key coordinating modes to develop resilience and creativity when facing a crisis, whether in the case of a crisis faced by an organization (The Lego Group) or a crisis faced by society (the Covid-19 pandemic). In the last part, Part six, as a conclusion, we derive an operational approach that managers can use to collaborate with innovation communities. This open management approach must learn to coordinate and dialogue with the organization's internal and external communities.

Part 1: Synthesis of the Theories on Innovation Communities

— In Chapter 1, Patrick Cohendet, Laurent Simon and Benoit Sarazin explain the dynamics of communities and the powerful source of ideas they represent. After describing the different types of communities and how they work, the authors highlight the importance of the "middleground," a required context for the community to develop. They show that this middleground is the channel through which the company can create a link with the community and access its creative potential. In order to establish this link, the company's managers must adopt a new approach. They have to be the knowledge gardener who provides the community with fertile ground for its development while accepting not to control the emergence of ideas. The managers can then count on an ideas reservoir that should be cultivated as needed while ensuring regular feedback to the community.

Part 2: The Dynamics of Communities of Innovation in Organizations

— Chapter 2 explores how Ubisoft has cultivated a strong relationship with multiple communities that infuse a strong dose of creativity into its various video game creation projects. Patrick Cohendet and Laurent Simon first highlight the role of communities of specialists in new professions in the video game industry. These specialists are in contact with the creative communities of the city of Montreal, with whom they exchange ideas in creative venues and during festivals and events that enliven the life of the city. Ubisoft also benefits from the communities of users of its products who contribute to their development. They are also in contact with communities of external specialists, such as the historians who helped develop the *Assassin's Creed* series. In order to establish a link with these communities and access their creative reservoir, Ubisoft's management has adopted a specific attitude that consists of generous action towards the communities, a discreet presence in their exchanges and continuous appreciation and recognition of the community's contributions.

— Chapter 3 describes the relationship between Salomon and the sports community. Benoit Sarazin and Jean-Yves Couput show how the brand supports the creation and development of communities such as the trail running community. Thanks to the contributions of the community's lead-users, the brand benefits from unexpected inspirations allowing Salomon to enrich its products and maintain an advantage over competitors. The authors depict how the middleground plays a key role in both the functioning of the community and the relationship between the brand and the community. They draw lessons about good practices to be used in community relations: delivering value authentically, being transparent in communication. They identify the pitfalls into which companies easily fall, such as one-way exploitation, and which provoke rejection from community members.

— Chapter 4 presents Schneider Electric communities of practice that gather employees who share a passion for same topics and projects. Louis Pierre Guillaume, Coline Delmas and Karine Goglio-Primard show how the company, becoming aware of the contribution of communities to increase collective intelligence, decided to promote and support the creation of these communities through the Community@work program. The authors explain the conditions of success of such a program. For example, they emphasize that measuring the value brought by the community can only be done through a satisfaction survey among members and not by measuring the concrete results coming from the communities. They demonstrate the importance of a network of community facilitators that allows community members to help each other by exchanging good practices and new ideas.

— Chapter 5 focuses on SEB's community of practice in innovation, a large French firm which includes 1,300 employees and 30 subcommunities. In this chapter, Lusine Arzumanyan, Charlotte Wieder and Claude Guittard highlight how it revolves around an annual event, the Innovation Forum, as well as around the facilitation action carried out in each subcommunity and orchestrated by the coordinator of the community of practice. They outline key success factors: the intrinsic motivation of the members, the circle of trust, the operating rules to which everyone adheres, the community managers' concern to be at the service of the members, the facilitation plan to set the pace, the culture of permanent feedback and the presence of an involved sponsor. They also highlight how these good practices challenge traditional management roles.

— Chapter 6 is dedicated to innovation communities at Schmidt Groupe. Tristan Cenier and Patrick Llerena relate the results of an experiment: the animation of a Creativ'café where volunteer employees generate innovative ideas using methods that differ from the traditional innovation process: no obligation of result, no agenda, short sessions and a "fun" atmosphere. They detail the process of these sessions, the methods used and present six ideas that came out of them. They show that a community emerges naturally from these events. They explain the obstacles encountered, whether in the facilitation of the sessions or in the difficulty for the bearer of an idea to accept that it can be rejected.

Part 3: Communities of Innovation in Industrial Ecosystems of Multiple Organizations

— Chapter 7 is dedicated to Renault's innovation community. Frédéric Touvard and Dominique Levent explain how this community was born out of the desire of a few individuals who felt the need to renew innovation practices. Relying on heterogeneous members, the community gathered them by cooptation, with representatives from some forty companies, the academic world, philosophers, historians, artists and sociologists. It follows established rituals such as plenary meetings punctuated by the offbeat and reflexive contributions of philosophers. It relies on doubt as a condition for the emergence of ideas and pushes for their concretization in prototypes. It has created mechanisms that encourage creative exchanges such as "distillation" and "boxing". The authors also report on the issues faced by the community: What is the future of the community? How can we resist its institutionalization and rigidification? How to react when its members form other competing communities and attract creative enthusiasts who are no longer available for the activities of the Renault community?

— Chapter 8 concentrates on ecosystem innovation and Michelin's innovation community. This community orchestrates around Challenge Bibendum, which takes place every two years, and the Open Lab, the structure that allows a hybrid community of different companies to work continuously on common projects. For Michelin, ecosystem innovation consists of bringing together companies interested in the evolution of mobility beyond its circle of suppliers or customers. Erik Grab details the Open Lab's objectives and the operating mode that enables it to achieve them. He specifies the rules that help create trust while preserving the confidentiality of projects. He shows how community members get motivated by a shared vision of the societal issues surrounding mobility.

Part 4: Mobilizing Communities of Innovation Through the Middleground

— Chapter 9 focuses on online communities, using the examples of the community of players of the online video game *Trackmania* created by the company Nadéo, or the community of Internet users who use Free's services. Guy Parmentier shows that brands stimulate the creativity of communities by opening up their products and designing them to encourage their appropriation, or even their diversion. They offer tools with which users can create product extensions and then integrate these creations into the product. In addition, brands establish a strong relationship with the leaders of the different communities involved and contribute to their animation. The author also discusses the challenges faced by the company in this collaboration, especially when the company, driven by its success, becomes a large, established organization.

— Chapter 10 describes how Hacking Health mobilizes a community of innovation through events. Hacking Health is a non-profit organization that offers heterogeneous groups the opportunity to create innovative solutions in the field of healthcare. Events bring together healthcare professionals, IT developers, designers, entrepreneurs and investors. They take the form of intense 48-hour hackathons, short sessions such as cafes and workshops, and longer events, the month-long Cooperathons. Karl-Emanuel Dionne, Luc Sirois and Hugues Boulenger unveil the keys to the success of hackathons in developing innovative initiatives that break with traditional practices that are often too institutionalized: the event's limited duration, the inspiring mission that unites participants, the momentum created by community membership, the animation of the event and the ability to create a sense of community, the structured management of creative process and the necessary follow-ups for project implementation.

— Chapter 11 explores how an external actor stimulates territorial innovation through the development of its middleground layer. Thanks to the case study of an open technology project in the field of ocular surgery, Karine Goglio-Primard, Odile de Saint Julien and Florence Crespin-Mazet show how a South-African organization (Afrikaner) contributes to fertilize plastic and ocular surgery fields in Delhi (India) through a well-structured process based on three stages: the identification of a fertile milieu for technology transfer; the structuration of its middleground through the creation of an incubator; and the professionalization of both middleground and underground layers to enhance territorial connectivity. The authors especially identify the use of a manifesto and a codebook; the creation and implementation of an incubator and the professionalization process of local actors grounded in entrepreneurial practice as key factors. Finally, they draw on managerial implications for policy makers.

Part 5: Communities of Innovation as Key Coordinating Modes to Develop Resilience and Creativity when Facing a Crisis

— Chapter 12 highlights the way firms might orchestrate tensions related to the collaboration with growing user communities for innovation. In this chapter, Émilie Ruiz, Romain Gandia and Sébastien Brion show how firms that harness mature user communities can balance control and autonomy and thereby maintain the innovation activity over time. Studying two firms from the entertaining field, the LEGO Group, leader of the toy industry, and Ankama, a major French firm from the videogame industry, the authors analyze both firms' user communities, strategies and attempts to deal with the focal tensions and orchestrate their communities. It appears that both formal and informal user community orchestration mechanisms allow firms to balance the control and autonomy tension when they collaborate with growing communities, such as the formalization of explicit key roles for some of these communities' users, the implementation of rewards or the development of specific auto-regulation rules.

— Chapter 13 analyzes the dynamics of communities in the context of crises such as the Covid pandemic. Zoe Masson and Guy Parmentier explain how the crisis pushes individuals to improvize and find solutions by developing multiple interactions and by mobilizing their creativity. In particular, they show how virtual communities are a favourable place for sharing the creativity of Internet users. With the development of the Internet, these communities develop a new form of socialization conducive to knowledge sharing and creativity. Thus, in response to the problems caused by confinement, new communities have emerged in many fields such as education, sports, politics, local life, research, engineering, etc. The authors emphasized that these communities seem to have their own characteristics that distinguish them from the traditional categories of virtual communities identified in the literature.

— Chapter 14 by Madanmohan Rao extends the scope of communities from innovation to resilience, particularly in light of the coronavirus pandemic. It begins by reviewing some of the literature on knowledge management, innovation, entrepreneurship and resilience, and shows how communities of practitioners can be key players in this regard. Based on insights from 25 organizations, it highlights emerging trends in the field of innovation communities, such as the growth of cross-sectoral and cross-organizational communities. During the pandemic crisis and in the post-COVID era, resilient communities will play a key role in sustaining and scaling the next waves of innovation. They will help organizations evolve and refine how to 'pivot, persist or pause' in their current strategies.

Part 6: Conclusion

— Chapter 15 summarizes the practical lessons learned from these case studies. Benoit Sarazin, Laurent Simon and Patrick Cohendet provide a guide for the manager seeking to benefit from the creative input of an innovation community. They identify the contributions and benefits that the company can expect from communities, depending on the type of community involved. They explain the needs of the members of a community, which the manager will have to take into account. They reveal that the company can establish a link by contributing to the middleground surrounding the community. Finally, they set out the conditions for success that will make the collaboration fruitful and allow organizations to take full advantage of their relationships with innovation communities.

Part 1

Synthesis of the Theories on Innovation Communities

© 2021 World Scientific Publishing Company
https://doi.org/10.1142/9789811234286_0001

Chapter 1

Communities of Innovation: A Synthesis

Benoit Sarazin, Patrick Cohendet and Laurent Simon

As a key issue for socio-economic development, innovation in the business world is currently undergoing a profound and irreversible upheaval. Until a few years ago, innovation processes were conducted in organizations according to well formalized methods, ensuring the framing and formatting of new ideas from the R&D laboratory to the market. These proven methods regulated in a coordinated manner the activities of formal groups within the company (functional departments such as methods offices, engineering, production management or marketing) involved in innovation processes structured in project teams and driven by a strict sequence of activities. As technological and business challenges become more complex, these traditional forms of innovation are no longer sufficient to provide the organization with the knowledge and ideas needed to develop new and appropriate solutions.

In this book, we support the idea that innovation will increasingly draw its main source from innovation communities. An innovation community is an informal group made up of either internal members of companies or external members (users of the products and services of these companies, informal virtual groups sharing a common interest, etc.). What is strongly underlined by all the cases described in this book is that these informal groups are increasingly playing the role of true active units in the innovation process, through which creative ideas emerge, are validated, tested and implemented. In concrete terms, the relationship with the communities goes through a structure called the middleground, which we describe in what follows.

I. Typology of Innovation Communities

The communities highlighted, both in the management literature and in managerial practices, are numerous: practice, user, virtual, interest and epistemic communities.

1. *Communities of practice*

Since the early 1990s, companies have seen communities of practice (Lave and Wenger, 1991) as one of the best ways to share knowledge, identify good practices, avoid repeating the same mistakes and seek common solutions. A practice community is made up of a group of members (usually from the same organization) engaged in the same practice — the practice of a trade, for example — and communicating regularly with each other on this area of knowledge (through mechanisms that can be very diverse: e-mails, face-to-face meetings, seminars, etc.). The different internal communities of specialists at Ubisoft's Montreal studio (game designers, 3D graphic designers, scriptwriters, etc.) detailed in Chapter 2 of the book illustrate this type of community of practice. Such a community can be seen as a coordination mechanism allowing its members to improve their individual skills, through the exchange and sharing of a common repertoire of resources that are built as the community's practice develops.

2. *User communities*

While members of communities of practice are characterized by their membership in a given organization, the work has shown that communities can also extend beyond the boundaries of organizations. Thus, communities of users of given products or services are very important potential sources of innovation (as von Hippel had already pointed out in a pioneering work in as early as 1986). Increasingly, user communities are developing around brands or products that develop knowledge beyond the boundaries of organizations. This is a very widespread phenomenon in the world of new technologies — the community of fans of Apple products, for example — digital entertainment — a community of players assembled around the experience of a video game, or even a community of fans, carrying high the knowledge associated with an artist or a sport... For example, in this book, the chapters on Salomon's innovation community (Chapter 3), Schneider (Chapter 4), and Seb (Chapter 5) illustrate the key role played by users (such as Salomon's trail running shoe lead-users, for example) in the company's innovation processes.

3. *Virtual communities*

The economic literature on communities also takes a particular look at the analysis of the functioning of virtual communities in connection with the development of the Internet, such as the hacker community in the context of the development of the Linux operating system. The development by Nadeo (Chapter 9) of a very active online community of video game players who create activity, content and innovations is a good illustration of this type of community.

4. *Communities of interest*

Emphasis has also been placed on communities of interest (Fischer, 2002) centered on members who share a common interest, such as certain groups coming together around a cause, such as families of Alzheimer's patients, or advocates for an endangered ecological site. They can also play a role in innovation for industrial groupings, as shown for example by the case of the different communities orchestrated by Michelin's Open Lab (Chapter 7).

5. *Epistemic communities*

Particular attention has also been paid to epistemic communities (Cowan *et al.*, 2000) which are focused on the deliberate production of new knowledge (e.g. scientific or artistic communities). Thus, in the artistic field, an epistemic community may include promoters of a particular style or movement, such as the Impressionists or the Surrealists. In science, they include defenders of the same theory, such as the advocates of general equilibrium in economics, or the supporters of the hypothesis of the origin of man in East Africa in paleontology.

II. Internal and External Communities

In the 1990s, only internal communities were recognized by companies. There are many examples of large groups that have explicitly stated the importance of these internal communities and have adapted their management to develop their potential. Whether it is the Learning Groups of Hewlett-Packard, the Family-Groups of Xerox, the Peer Groups of British Petroleum, the Knowledge Networks of IBM Global Services, the internal KM club of EDF-GDF, or the communities of practice of Caterpillar, the reference is always the same: these firms very quickly recognized the considerable potential contribution to their performance of these informal groups that bring together members of the organization around a particular area of knowledge. Three chapters of this book (Chapter 8 on Schneider Electric, Chapter 9 on SEB, Chapter 10 on Schmidt) present internal communities.

Practitioners and researchers then recognized that certain communities could also be deployed between inside and outside the organization, or even entirely outside, thus posing challenges for interaction with firms. As early as the 1990s, Toyota developed interaction platforms and practices with its main suppliers to facilitate the mobilization of engineering communities, each community bringing together both Toyota engineers and engineers from partner companies. As Dyer and Nobeoka (2000) point out, in this context, "the driver of innovation is no longer the individual firm, but the network." At the beginning of the 2000s, the multinational Procter & Gamble experienced a major innovation crisis and sought to stimulate and enrich its projects, particularly its innovations in applied chemistry. As a first step, the firm made a systematic effort to connect with the chemical researchers working at its suppliers' sites, facilitating community exchanges via electronic platforms. Then, in a second phase, it created, in particular with the pharmaceutical company Elly Lilly and NASA, a shared electronic platform on which thematic idea competitions were organized, which made it possible to connect to various communities of experts from different scientific and technological fields. These two initiatives had a recognized positive impact both on the diversity and speed of innovative projects and on their cost and profitability (Huston and Sakkab, 2006). Through these initiatives, the communities involved were able to take the lead in the development of innovative projects.

The project is designed to enhance their value by responding to complex challenges, thereby continuing their knowledge development project.

Finally, the various communities were able to gain a better understanding of other communities and thus expand their knowledge network. Several chapters of this book (Chapter 3 on Solomon, Chapter 4 on Schneider, Chapter 5 on online communities) present communities whose members are external to the organization: the users of the company's products. Other chapters (Chapter 6 on The Schmidt Groupe, Chapter 7 on Renault) describe "mixed" communities, i.e. the community includes members both inside and outside the company.

The formation of a community can also be totally autonomous and purely virtual, outside any prior organizational framework. This is the case of the hacker community in the context of the development of the Linux operating system. These hackers formed a community spontaneously, without the mobilization of a firm or a formal organization. They were professional computer scientists and enthusiastic hobbyists who, frustrated by the limitations of the Unix operating system, began programming according to their knowledge as developers and their user experiences in order to facilitate their own work. They naturally shared the code they had designed with other passionate computer scientists who were dealing with the same problems. The immediate benefit of this sharing was to increase the quality of the code and the efficiency of their system, as well as to simplify their work. Indeed, the person with whom they had shared their code detected bugs and errors, corrected them and transmitted the corrections to their author. A set of ad hoc contributions developed around this common project, which were aggregated and validated by the community of developers themselves. The commercial exploitation of Linux came later and remains peripheral to the community, which is still very much in the forefront of the development process (Lessig, 2002).

III. Our Proposal: To Group all Forms of Innovation Communities Under the Same Title

In this book, we do not attempt to go into this clear and now well-established typology of the various communities. On the contrary, we propose to group all these informal groups (communities of practice, epistemic, virtual, etc.) together under the expression "communities of innovation". The reason for this bias is the following: until recently, all these communities remained marginal in relation to the innovation dynamics of the firm. Studies have focused on examining the behaviour of individual communities in isolation. And even if one recognized the occasional contribution of a community within the framework of a given innovative activity, the innovation process as a whole remained the prerogative of traditional R&D departments, of methods already in use or of marketing through formal and controlled processes. The major change that is taking place is that today, in many organizations, communities are no longer limited to peripheral devices.

The company's business units have become central active units that serve to generate and validate new ideas that form the basis for original products and services. As Brown and Duguid (1991) predicted, the firm is becoming a true "collective of communities", where the capacity for innovation increasingly relies on the combined contributions of the various communities. From this perspective, the aim of this book is to shed light, using multiple examples, on the proactive and fundamental role of communities in the new innovation practices of organizations.

Box 1: Note

Although we include all types of communities in the single category of innovation communities, it is useful to differentiate between internal, external and mixed communities when specifying the company's action towards the communities. Indeed, operational issues are not the same in each of these categories. This is what we do in Chapter 15 on "The practice of communities".

IV. The Dynamics of Innovation Communities

1. *The functioning of innovation communities*

In each of these communities, the behaviour of the members is characterized by their shared passion for a given subject. This passion translates into voluntary commitment to building, exchanging and sharing knowledge with other community members. These communities are therefore essentially made up of a core group of enthusiasts, which attracts participants with varying levels of interest. They are generally autonomous and focused on the production and sharing of knowledge and experience to deepen understanding of a topic or improve a skill.

Communities create strong bonds among their members. These bonds are based on the passion and commitment of each member to a common goal or practice. Interactions between members of a community are governed by relationships of trust based on respect for norms (some of which are community-specific). Trust can be measured by the fact that, in the face of unforeseen events, the behaviour of individuals is guided by respect for the norms established within the community and not by contractual patterns.

Repetitive interactions within communities significantly reduce opportunistic behaviours that are replaced by "routines" (Nelson and Winter, 1982), norms of cooperation and reputational mechanisms. For example, Lerner and Tirole (2001) show that individuals within their communities are motivated by the reputation they gain among their peers. The activity that takes place there is carried out without the direct control of an explicit hierarchy that would seek to monitor compliance with procedures or the quality of the work provided. The concepts of contract and incentive pay are therefore secondary, if not totally absent.

Community members' awareness of belonging to a community is built around activities which are commonly understood and continually renegotiated between its members. Thus, for example, the small community of surfers in Hawaii described by von Hippel (1986) revolutionized the surfboard, its shapes, its materials, in the late 1970s, as each member shared his impressions, sensations and new ideas with the community after a practice session. In this sense, a member feeds his community with the experience he acquires in his daily practice and, in return, can rely on the resource pool — the accumulated, shared and continually sought-after knowledge — that the community maintains to carry out its activity for itself or for an organization. There is no authority guiding practice: it is a shared commitment that binds the members of the community of practice into a coherent social whole.

The validation of the knowledge and ideas produced is carried out primarily within the community. In addition, knowledge emanating from outside a community (the hierarchical structure of the company, for example) is evaluated, examined, reinterpreted (sometimes resulting in "creative deviations") by the community before being assimilated.

Communities do not have clear boundaries. They are not linked to an organizational chart and can easily cross the boundaries of the business. It is difficult to determine where the community begins and ends: a member belongs to the community by a spontaneous act that is not recorded anywhere and can be interrupted as soon as the member is no longer interested in the subject carried by the community.

V. The Life Cycle of Communities

All communities have life cycles, from their emergence to their maturity and, eventually, to their "death". In a much-cited article on the formation and evolution of communities at IBM Global Networks, Gongla and Rizzuto (2001) highlight the existence of community life cycles, emphasizing that the sustainability of a community is never assured. In particular, a community may disappear for two opposing reasons: either because the creative spirit that animated it and that led the passion of its members to invest in exchanges and interactions within the community has finally died down ("there is no more grain to grind"), or, on the contrary, because the community has worked so well to develop new concepts or products that it is time to "institutionalize" it. Using the case of communities within a given organization (IBM Global Networks), the authors show how the management, having seen the success of the preparatory work done by a community, decides to create within the organization a new functional department, or service, or product line that "institutionalizes" the functioning of the community. IBM Global Networks' growth dynamic is thus based on the remarkable ability to continuously foster the emergence and development of diverse communities within the organization. Many of these communities "do not take" or become depleted, but some reach remarkable stages of development that lead the organization to institutionalize them through the creation of new services or departments that create value for the organization. These perspectives highlight the fact that a community can be seen as a "potential re-serve for the future of the organization".

These phenomena of institutionalization of communities are not only observed within a given organization. As an example, Rabeharisoa and Callon (1999) analyze the case, from the 1980s, of the community of parents of children with cystic fibrosis, an "orphan" disease whose rarity explained the lack of interest shown by researchers, physicians and pharmaceutical companies. This situation of abandonment led the parents of these children to form a real community of practice, which was able to gather and accumulate, based on patient observations, useful knowledge to improve the situation of the children. This knowledge was then able to begin to be useful to the research community in order to improve and pursue their research in depth. Faced with the lack of money to significantly finance the research, this community of parents was able to convince the leaders of French television to organize the first "Telethon" to collect donations in order to guarantee the continuation of the research. The remarkable point the authors make is that, at the end of the Telethon, when the director of the television channel, delighted with the unexpected success of the event, asks the parents: "Who do I give the cheque to?", the parents have no other solution than to join together in a structured association (with a president, a treasurer, etc.). This association (the AFM) has created a vast movement of solidarity and financial support for research and has subsequently invented an original model in which lay people dialogue with specialists, without losing sight of the objectives assigned to science. It should be noted here that the community finally "institutionalized" itself in the form of a formal organization.

Naturally, many communities wish to avoid both the pitfalls of breathlessness and institutionalization. This is the case for most of the communities analyzed in this book. As an example, the chapter on the Renault community (Chapter 7), which shows the different phases in the evolution of this community, clearly reveals in filigree that this community wishes to keep its dynamism and does not envisage being institutionalized. The chapter highlights the efforts made by those who coordinate the community's operations to keep the community active, to stimulate passion and collective sharing, and to nourish members with new challenges and perspectives.

Regardless of the nature of a community's life cycle, a common thread running through all communities is the willingness of members to structure their community in a way that is meaningful to their cognitive effort and to specify their joint action program. These shared goals are usually expressed in a manifesto and a codebook. These concepts of manifesto and codebook (detailed in what follows) have been specifically developed in the study of epistemic communities, but we consider them useful to represent the creative process of any given informal group.

VI. The Manifesto and the Codebook: Key Structuring Elements of a Community

The work on epistemic communities has shown that, despite their informal nature, these communities have two structuring elements: the manifesto and the codebook. First of all, their members agree on the elaboration of a "manifesto", a real program of action (like a new artistic or scientific movement) that expresses how this new movement breaks old rules and proposes new ones, based on specific values. We can think here of the Cubism or Surrealism manifesto, or the new circus manifesto proposed in the early 1980s by the Cirque du Soleil, which is built around four rules that break with the old-style circus rules (“no animals”, “no curtain”, “original costumes”, “music on stage”).

Once the manifesto has been explained to guide the creative action of the community members, they will be able to engage in a second stage of the creative process: the elaboration of a “codebook”. The codebook is a dictionary and grammar book that complements the manifesto. In a way, it is the “user’s manual” for the implementation of the new knowledge promoted by the community. For example, the Cirque du Soleil codebook is the set of rules, practices, modes of use and new vocabulary accumulated by the community that was created around Guy Laliberté and the colleagues who founded the manifesto to produce the first shows of the new circus in the 1980s. Little by little, the codebook is stabilizing and represents what can be passed on to a newcomer.

VII. Innovation Communities Bring Major Benefits to Businesses

The work of Lave and Wenger, as well as Brown and Duguid, was the first to demonstrate the exceptional effectiveness of the communities in comparison with traditional processes. They concerned “practice communities”: informal groups that focus on the exchange of best practices between people doing the same work. They illustrate four key aspects of the effectiveness of innovation communities.

The first is that communities create, share and store knowledge much more effectively than do the formal structures of an organization. The case of the Xerox photocopier repairers is perhaps the best known and most representative example of how a community of practice works. In the case of Xerox, the work of a photocopier repairman in the early 1980s could be described as a continuous improvization in a network of relationships between customers, machines and other repairers (Orr, 1990). Each repairer usually operated alone, on customers whose machines (at the time large photocopiers occupying an entire room) were broken down. However, while the repairers (the “reps” in their jar) worked largely autonomously, their major problem was that the detailed, codified manuals produced by Xerox’s hierarchical organization to solve customer repair problems were largely useless, inappropriate and unusable. So, together they gradually formed a true community in which experiences were exchanged, good "tricks" and "tinkering" that worked were shared, primarily through "war stories" that allowed for collective learning and the sharing of knowledge useful to their business. This resulted in a repertoire of practical operational knowledge that was far more effective than the (mostly unusable) recipes in the manual developed by the hierarchy. Over time, the repairer community even established its "rituals" and shared social experiences, such as the "initiation ceremony" where the newcomer (the newly hired repairer) had to burn (in secret, in front of the other repairers) the manual given to him by the hierarchy.

The second aspect is the high quality of the ideas generated by the community. Even if community members are not interested in finding a market for their ideas, they are still passionate about sharing their ideas with other members of the community. They therefore enter into a process of "seduction" and explaining their initial ideas to their peers. This interaction within the community, fuelled by feedback and reactions from community members, has the effect of making the initial creative idea stronger, more intelligible and more valuable.

The third aspect is that a community can open up new perspectives which, through "contagion" with other communities in the organization, lead to the formation of radical innovations. If we take the example of Xerox, the repair community also did not operate in isolation: through their practice and their constant interaction with customers and technicians, the reps have also forged a representation of what the new generations of photocopiers should be, given the constraints of technology and customer expectations. Through constant interaction with customers, Xerox reps were the first to see the need to move to a new generation of products by replacing large photocopiers with small desktop photocopiers. The reps were not just repairing machines: through their commitment to their practice, what emerged was the co-production with other communities within the company, particularly in R&D and marketing, of a shared vision and representation of the evolution of photocopiers. This shared vision eventually convinced Xerox management to pioneer small office photocopiers, making the company the world leader in this vast market.

The fourth aspect is the reduction of costs brought about by the communities in the creation, sharing and preservation of knowledge. As knowledge grows and becomes more complex, traditional hierarchical structures are finding it increasingly difficult to integrate and develop patches of specialized knowledge. They are reluctant to bear some of the fixed costs associated with the processes of knowledge creation and maintenance. Indeed, the construction and maintenance of a given body of knowledge requires the patient and costly development of specific models, grammar, languages, codes or rules. For example, for a firm such as Philips today, it seems more profitable to invest in active and dynamic connections with different communities of experts and users than to maintain only classical R&D poles in the form of scientific and technological research laboratories (Blau, 2007). In the case of Ubisoft's studio in Montreal (Chapter 2), the firm's delegation of part of its knowledge development and memorization work to internal and external communities also demonstrated remarkable efficiency gains. From this point of view, informal communities offer, through the voluntary and "free" commitment of their members, the advantage of being able to generate and consolidate parcels of specialized knowledge at low cost. By acting as the active memory of the organization, these communities are particularly able to bear some of the fixed costs associated with the processes of knowledge creation and maintenance.

The benefits for managers of using the concept of community to foster knowledge sharing are numerous: increased staff performance and know-how; increased productivity; greater collaboration; improved efficiency; operational effectiveness and interest in work. For example, managers are frequently faced with the well-known problem of retaining the memory of the knowledge acquired during a project when it is completed (and the project team members have returned to their home departments). Communities offer them an efficient and inexpensive solution: through regular exchanges as long as the community is active, the memory of the knowledge acquired and accumulated is preserved (at negligible cost). Lesser and Storck, who conducted a study for IBM in 2001, report, for example, that in the case of International Data Corporation, the mobilization of different communities of practice existing within the firm avoided "reinventing the wheel" on an ongoing basis, saving up to the equivalent of US$5,500 per employee in lost time and inefficiency. The community can thus be seen as a "slack" of knowledge and creative ideas that can be drawn upon to contribute to innovation performance at any given time.

The contribution of communities is all the more relevant since innovation must increasingly address systemic issues that call for diverse and interdisciplinary knowledge: uses and their variety, social impacts, environmental impacts, ethics, symbolic dimensions, etc. The contribution of communities is all the more relevant since innovation must increasingly address systemic issues that call for diverse and interdisciplinary knowledge: uses and their variety, social impacts, environmental impacts, ethics, symbolic dimensions, etc. The contribution of communities is all the more relevant since innovation must increasingly address systemic issues that call for diverse and interdisciplinary knowledge: uses and their variety, social impacts, environmental impacts, ethics, symbolic dimensions, etc. R&D or marketing teams are ill-equipped to integrate these issues into their work. Yet it is the external communities that carry different pieces of the puzzle of understanding these new issues and problems. By connecting with these communities in real time, the company greatly improves its ability to adequately address these systemic issues.

VIII. The Middleground: A Generative Platform Between the Organization and the Communities

1. *The description of the middleground*

Given the source of creativity provided by communities, the business community has an interest in the existence and development of communities in areas of interest to it. Once these communities exist, it will want to establish a strong relationship to collect the "creative honey" and nurture the organizations' formal innovative processes. However, there are two problems with this approach.

First, a community is by definition an entity that wants to remain spontaneous, informal and as unorganized as possible. It is a truly autonomous group based on a principle of voluntary membership based on the sharing of a certain number of values, norms, common cognitive interests or a common practice. As a result, the manager cannot rely on the traditional mechanisms of the organization, decided and set up under the authority of the hierarchy (functional groups and project teams in particular). He cannot decide to create a community around a theme that interests him and then lead it. An innovation community is difficult to decree. It is not managed from the outside but is self-organizing from the inside. Any attempt by the hierarchy to "intrude" directly into the internal workings of a community resulting from a process of self-organization would be doomed to failure. For example, this type of failure was experienced by Nortel when, in the late 1990s, the firm abruptly and massively bought out companies on the Wall Street market to make up for its technological backwardness in the wireless Internet (Amin and Cohendet, 2004). In the acquired companies, seeing the brutal intrusion of a hierarchy, the members of the scientific and engineering communities enforced ethical clauses of no prior consultation and went in particular to the competitor Cisco (which did not intrude into the functioning of the communities). Nortel found itself with empty shells that it had bought at a high price just before the dot.com bubble burst. This mistake precipitated the fall of the company.

On the other hand, it is necessary to channel the relationship with the members of the community. Indeed, it is illusory to spend time developing hypothetical and unreliable interactions with the anonymous crowd of individuals and creative talents that make up the community. Some direct actions are possible (e.g. via crowdsourcing techniques, as in the case of the Fiat Mio concept project produced with 17,000 idea givers, cf. Saldahna *et al.*, 2014), but they are generally expensive, very random and most often not very effective.

The solution for the company to interact with communities is to use an intermediate stratum between the organization and the communities that we call the “middleground”. The middleground is a notion highlighted in the work on creative territories (Amin and Cohendet, 2004; Cohendet *et al.*, 2010). It is conceived as a physical or virtual context between the formal structures of the organization we call the “upperground” and the set of creative “talents” of the communities we call the “underground”. The latter can be individuals from outside the organization (users, specialists in a field of competence of the organization, etc.), as well as members of the organization developing ideas not yet taken into account by the organization (for example, a Ubisoft employee developing new video game ideas and “tinkering” in the evening in his garage).

Research shows that the middleground is a common good, co-created by members of the community (the underground) and the organizations that run it (the upperground). The middleground is the necessary context for the community to develop, it is characterized by 4 mechanisms: places, spaces, events and projects (Cohendet *et al.*, 2010). These fundamental components of the middleground (see Fig. 1) have the following properties:

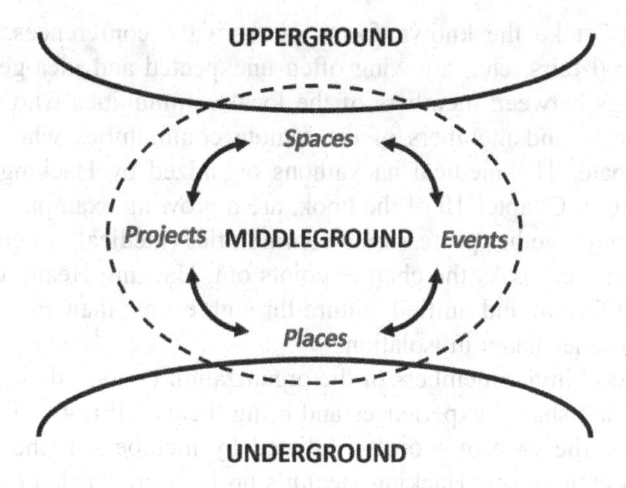

Figure 1: The 4 key mechanisms of the middleground: Spaces, projects, events, places.

— "Places" are physical spaces for socializing experiences (third-party places, fablabs, "places of encounter") that are used for socializing (e.g. co-design rooms, cafés, etc.), as well as possibly virtual spaces for sharing and combining knowledge (forums, wikis, etc.). One of the places mentioned in this book is Ubisoft Montreal's famous terrace, a real platform for meetings between the studio's different communities of expertise, but also for improbable encounters (for example, between guest artists and studio designers). The squares are thus not simply places for socializing between "usual suspects", but also, as was the Agora in Athens, places where unexpected encounters with a passing "stranger" can trigger creative ideas.
— "Spaces" are cognitive spaces that favor within the middleground the construction of ideas as well as their diffusion and outreach to the outside world. For example, the cubists before the first war in Paris were at the origin of the construction of a new school of painting (space) which spread its ideas throughout the world (notably through a famous manifesto). When top sportsmen and women, the lead-users of Salomon's products (Chapter 3), invented Alpine Running, a new sport halfway between trail running and mountaineering, they created a new space. This constructive exchange of ideas can also take the form of launching new styles or new trends that have an influence and an echo outside.

— "Events" take the known forms of festivals, conferences, forums, industrial fairs, etc., allowing often unexpected and idea-generating meetings between members of the local communities who organize the events and members of the remote communities who come to participate. The medical hackathons organized by Hacking Health, featured in Chapter 10 of the book, are a growing example of events that bring together professional communities (medical, programmers, designers, etc.). As the chapter points out, Hacking Health creates a more informal and unified culture through events than each of these communities taken in isolation.
— "Projects" invite members of the organization to align their common vision and shared experiences and bring them to fruition. This is for example the case of projects initiated by members of the Hacking Health community. Hacking Health's projects are carried out in two phases: first, Hacking Health organizes events, hackathons mentioned earlier, where small groups of volunteers made up of health professionals, developers and designers work together in a limited time frame to develop projects to solve health problems. At the end of the hackathon, the most promising projects are selected. Their development can be supported by one of Hacking Health's partner companies, such as the Business Development Bank of Canada or Desjardins. Thus, in this context, the events (hackathons) promote the emergence of federative projects that develop over the long term and contribute to increasing a vision shared by the members of Hacking Health. Similarly, in the case of Ubisoft's Montreal studio, for example, the organization of an independent game creation competition can be likened to a "project" component of the studio's middleground: the event federates the efforts of amateur developers from the underground to launch projects that will be presented on the day of the competition. By giving them the opportunity to create, the company ensures a strong link with the community and detects emerging outside talents and new ideas. Projects can also appear independently of the events. This is the case of Decathlon Cocreation's crowdsourcing platform where users can submit ideas at any time that can become projects if they are approved by the community members.

The middleground will be all the richer, more dynamic and creative as these four components are simultaneously activated. Their simultaneity ensures permanent exchanges, not only between the members of a given community but also between the different communities. Indeed, the middleground also plays a role of transversality and connection between the various communities that are present there. All of these interactions promote the circulation and ongoing cross-formation of knowledge, ideas and projects, which are at the source of the generative dynamic of the middleground and its growing interest in corporate strategy.

2. *The benefits of the middleground*

The middleground brings two main benefits to the community (the underground) and the organizations (the upperground) that surround it:

— To provide the necessary fertile soil for the community to grow.
— Creating the climate of trust necessary to transfer ideas originating from the community to the upperground.
The middleground also plays a key role in all phases of the community's evolution:
— Emergence of the community: It is when individuals passionate about the same subject meet in an informal place (square) and confront their ideas with other currents of thought in cognitive spaces (spaces) that they build the manifesto and found the community. It is in informal places (squares) that they evangelize their ideas to other passionate people and recruit them to join the community. In the course of the exchanges, they make the manifesto concrete and practical by creating the codebook that defines the community's dictionary, its grammar and its specific practices. As such, we can say that the manifesto and the codebook, even if they are not formally part of the middleground, are fruits that could neither have taken shape nor matured without the fertile soil of the middleground.
— Functioning of the community: Once the community is formed, the middleground nourishes the dynamics of the community. It provides the means for members to meet in informal meeting places (squares), it gives rhythm to its activities and creates strong moments for members in events, it allows members to concretize their collaboration in projects and it gives them the means to broaden their thinking by confronting themselves with other currents of thought in cognitive spaces (spaces).
— Challenging the community: Community activities and knowledge are constantly being renegotiated between community members as

members make new contributions. This renegotiation is made possible and fluid through the exchanges between members that the middleground makes possible, in informal spaces of renegotiation, cognitive spaces, events and projects. In particular, the middleground is an essential asset when community leaders question the very relevance of the community, which can lead either to its rebirth or its death.

IX. Build Common Trust in Order to Transfer Ideas from the Community to the Upperground

The middleground allows a company or any formal organization to create a relationship of trust with the community and to counter the fear of its members to be "recuperated" by a hierarchy. Indeed, the company can show its goodwill by getting involved in the middleground, for example by providing the necessary means so that informal places, cognitive spaces, events and projects can exist. The organization thus adopts the principles of community, the first rule of which is sharing among peers. It is then perceived by the community as having an altruistic attitude, because it gives to others without expecting any direct return. It shows that it acts without interfering in the life of the community.

Once trust is established, the common ground allows for a win–win exchange between the organization and the community. Indeed, the contacts established thanks to the middleground allow the leaders of the organization to capture the ideas of the community which they find interesting. This happens when these managers meet community members in informal meeting places or at events. These meetings are extended when managers discuss projects undertaken by the community or participate in informal exchanges in cognitive spaces. The managers can then, with the agreement of their authors, achieve what the community cannot do: integrate these ideas into the organization's innovation process in order to transform them into concrete and competitive achievements on the market. They can then share the results of the innovations produced with the community and initiate a new innovation process stimulating a new cycle of creativity in the community. Indeed, on the basis of the innovations created, the company will be able to propose to the community to search for other ideas to further develop these innovations. It will also be able to submit to the community the difficult questions that have arisen during the development of the innovation. This feedback to the community is part of the win–win exchange between the organization and the community.

In short, the middleground is called upon to play a central role in opening up and connecting the organization to the ecosystem of its communities, which is now becoming one of the major challenges of corporate strategy. It is the real lung that enables the company to exchange with its external environment in an increasingly open universe.

X. Requiring a New Attitude for the Top Management

1. *Becoming a "knowledge gardener"*

In order to create a relationship with the community through the middleground, the manager must adopt a new attitude, one that contrasts with the processes of hierarchical organizations. It is a subtle and delicate art of management on the part of the hierarchy: Anglo-Saxon authors use the expression "to harness communities" to express this particular aptitude, which would resonate in French with the idea of respectful taming. The manager is called to become a "gardener of knowledge" who prepares fertile ground for communities to flourish. Even if he cannot dictate the actions of the community, he can know, support and even protect a developing community by contributing to the middleground, whether it is located inside or outside the company. He must be careful that the community does not perceive his action as interference that would lead to rejection. He must refrain from interfering in the way the community is governed. The organization must provide the resources without waiting for an immediately measurable result: the results will come naturally in informal exchanges between the community and the organization. These are the conditions for a complicity based on trust between the formal management of the organization and the community. Concrete actions that the manager can take to contribute to the middleground are described in Chapter 15 on community practice.

XI. The Case of Communities Designed by the Hierarchy

The difficulty of the relationship between the hierarchy and the community is visible in the case of hierarchy-led communities. The problem for business is that a community of innovation is difficult to decree and establish. Indeed, in the pure conception of the notion of community that prevailed prior to the introduction of the concept in the 1990s, trying to build a community "top-down" is a priori impossible: the very definition of the notion of community refers to a "voluntary and responsible adherence to a community".

The "membership" principle is a fundamental principle of community building. But the managerial interest in the notion of community has become such that many companies have tried to "force the concept" and have tried to gradually bring it into their vision and practice. This "conceptual diversion" has naturally come at the price of certain adaptations or reinterpretations and sometimes even at the price of certain slip-ups or disappointments. What is remarkable, however, is that the idea that companies could "pilot" communities of practice has gradually been reinforced. It is clear that today some companies are calling for the establishment of communities of practice that are led (by management). However, the characteristics, modes of implementation and limits of managed communities, which differ in many respects from spontaneous communities (which emerge autonomously among members sharing a common passion), remain largely to be clarified, analyzed and categorized. Thus, for example, the functioning and activation of a pilot community depends on the (necessarily costly) animation efforts of the hierarchical leadership: it is up to the hierarchy to ensure that the essential tasks of animation and coordination of the community are properly carried out. This role is particularly important in the early stages of community building. On the other hand, as the community develops and begins to be able to capitalize on the knowledge produced and exchanged, it is important that the hierarchy be able to gradually break down to allow the community mechanisms to develop fully. Chapter 4, on communities of practice at Schneider Electric, provides a concrete example of a company that has successfully implemented pilot communities.

XII. Broaden the Vision of the Organization

The manager must also broaden his vision of the organization. One of the main difficulties in analyzing innovation communities is that the division of the communities does not overlap with the division of the inhabited organizational structures. A community is generally not visible on an organization chart and therefore cannot be considered as a simple sub-category of the classic organizational decompositions of the company. While it may happen that a given community is composed of members from the same hierarchical division (functional department or project team), most communities are "cross-organizational" and include members from different functional departments or hierarchical teams. These changes pose a challenge to managers who must learn to "read" their own organization as a set of living knowledge bases that circulate and interact across and below titles, functions and job descriptions. In the case of Cirque du Soleil, for example, being able to identify who, across departments and functions, would be more inclined towards a streamlined or baroque style, allows for the assembly of interdisciplinary creative teams that have a strong common orientation and will be better able to collaborate on truly innovative projects.

XIII. The Principle of Reciprocity ("Give and Take")

Regardless of the type of innovation community considered, the manager must apply a principle of reciprocity of the "give and take" type ensuring a form of complicity based on trust between the formal management of the organization and the activities of the community. Indeed, if a firm exploits the contributions of a community without returning to it, its members will turn away and avoid engaging in exchanges in the future, thus closing a potential source of innovation. The toy firm LEGO has understood this and presents an emblematic case, investing for nearly 10 years in the implementation and gradual adjustment of a platform for capturing the ideas of the AFOL community (Adult Fans of Lego, expert fans of the brand, who create new models from its products). For example, LEGO has designed a system where ideas from community members are first evaluated by other members, before LEGO teams intervene to consider transforming them into a product. In this example, the goal is to not only feed the firm with new ideas by capturing them from expert fans, but also to enable the fan community to interact and to more easily identify its contributors, keeping the community active and creative. In addition, in the relationship between fans and the firm, the community expects the firm to listen and take note of the fans' proposals and to commit to their achievements. In this sense, dialogue between the firm and the community inspires the firm, reduces its risks and ensures a precise and cost-effective response to the community's needs (Antorini *et al.*, 2012; Schlagwein *et al.*, 2014).

XIV. Moving Beyond Organizations

Management must accept that exchanges go beyond the perimeter of organizations. For a given organization, one of the remarkable aspects of communities is that its members can regularly exchange knowledge with other members belonging to other institutions, other organizations, including even competing organizations. For example, the existence in Montreal of an independent association of video game developer employees — which acts as a middleground — allows regular meetings between employees of competing firms for activities to share good practices, which in turn contribute to the whole "cluster". The advantage is that communities can play the role of capturing external knowledge that is indispensable to the organization; but there can also be disadvantages related to the risk of strategic knowledge leaks to the outside world. In the case of Montreal game developers, participation in the association is encouraged by the firms, but issues of confidentiality of the firms' information are also often raised.

XV. Conclusion

All the chapters of this book converge to underline that the creative power of innovation communities is a new strategic resource for business. But all the cases analyzed also clearly show that capturing the creative potential of communities is not a matter of course. What is at stake is a real challenge to managerial practice. Managers cannot access the creative ideas of the community with the same transactional approach they use with their partners or suppliers. No contract, no formal cooperation agreement, no participation, no authoritarian hierarchical order to allocate work can be conceived between the organization and any community. Such an approach would be rejected by the community. Management practice must therefore change radically: the traditional manager, who focuses on developing a plan of action long in advance and seeks to dictate the activities of the members of the organization, must give way to a "gardener" manager. He must carefully watch over his links with communities and listen to the weak signals coming from these informal groups, like a gardener who takes care of his garden to obtain its fruits. The manager must first create a relationship of trust with the community members by contributing to the middleground, that is to say all the spaces and activities that allow the community to flourish and develop. He must then "enact", i.e. support the projects of the community that interest him and give them the necessary support to enable them to develop, without being able to interfere directly. He will have to agree to broaden his vision of organization because communities have no borders. He will have to question the modes of industrial property. Then he will be able to create a win–win relationship with the community.

These new managerial practices, however, do not mean that all areas of business management are the responsibility of a full-time manager-gardener. Entire sections of the company still have to be managed according to traditional patterns, particularly in the central phases of innovation processes, which are still managed by standard operating procedures, by the necessary stage-gate requirements, by strict control of suppliers and by the respect of commitments with the company's partners. What emerges is the fundamentally "dual" character of the manager of tomorrow, capable on the one hand of playing the role of community gardener, and on the other hand of coordinating the formal sequences of the innovation processes with the utmost efficiency.

In this new role, one of the main challenges facing managers is to know how to articulate the activities of different communities in a coherent and shared vision. While many of the reflections in this book have rightly focused on the interactions between the company hierarchy and a given community, much remains to be said about the links between different communities. These links are not obvious. Not only can discrete communities have very little interaction, but there are also risks of withdrawal, non-communication and even conflict between them. Thus, while a manager can hardly influence the work of a given community, he or she can make a strong contribution to facilitating the establishment and harmonious functioning of interactions between communities, and thus mobilize their potential in the service of the organization's development and sustainability. This managerial perspective opens up new and very important avenues for analysis and reflection that would justify an exciting follow-up to this first book.

References

Amin, A. and Cohendet, P. (2004). *Architectures of Knowledge.* Oxford University Press.

Antorini, Y. M., Muñiz Jr, A. M., and Askildsen, T. (2012). Collaborating with customer communities: Lessons from the LEGO Group. *MIT Sloan Management Review*, *53*(3), 73.

Baldwin, C., Hienerth, C., and Von Hippel, E. (2006). How user innovations become commercial products: A theoretical investigation and case study. *Research Policy*, *35*(9), 1291–1313.

Brown, J. S. and Duguid, P. (1991). Organizational learning and communities-of-practice: Toward a unified view of working, learning, and innovation. *Organization Science*, *2*(1), 40–57.

Cohendet, P., Grandadam, D., and Simon, L. (2010). Montréal, ville créative: diversités et proximités. *L'économie de la connaissance et ses territoires. T. Paris and P. Veltz. Paris, Hermann.*

Cowan, R., David, P. A., and Foray, D. (2000). The explicit economics of knowledge codification and tacitness. *Industrial and Corporate Change*, *9*(2), 211–253.

Dyer, J. H. and Nobeoka, K. (2000). Creating and managing a high-performance knowledge-sharing network: The Toyota case. *Strategic Management Journal*, *21*(3), 345–367.

Fischer, G. (2002). Learning through the Interaction of Multiple Knowledge Systems. *Working paper Center for LifeLong Learning & Design (L3D)*, Department of Computer Science and Institute of Cognitive Science, University of Colorado, http://www.cs.colorado.edu/~gerhard.

Huston, L. and Sakkab, N. (2006). Connect and develop. *Harvard Business Review*, *84*(3), 58–66.

Lave, J. and Wenger, E. (1991). *Situated Learning: Legitimate Peripheral Participation.* Cambridge University Press.

Lesser, E. L. and Storck, J. (2001). Communities of practice and organizational performance. *IBM Systems Journal*, *40*(4), 831.

Lessig, L. (2002). *The Future of Ideas: The Fate of the Commons in a Connected World.* Vintage.

Orr, J. E. (1996). *Talking About Machines: An Ethnography of a Modern Job.* Cornell University Press.

Schlagwein, D. and Bjørn-Andersen, N. (2014). Organizational learning with crowd-sourcing: The revelatory case of LEGO. *Journal of the Association for Information Systems*, *15*(11), 754.

Von Hippel, E. (1986). Lead users: A source of novel product concepts. *Management Science*, *32*(7), 791–805.

Part 2

The Dynamics of Communities of Innovation in Organizations

© 2021 World Scientific Publishing Company
https://doi.org/10.1142/9789811234286_0002

Chapter 2

Communities of Innovation at the Ubisoft Montréal's Studio

Patrick Cohendet and Laurent Simon

With over 2,000 employees, Ubisoft Montreal's studio is the largest video game development office in the world. Established in 1997 by the French-owned multinational group Ubisoft (one of the world's leading video game developers and publishers), the studio quickly became a creative flagship. It successfully launched many blockbuster games (over 5 million units sold), which became powerful brands for series development on consoles and other platforms (e.g. *Prince of Persia, Rainbow Six, Splinter Cell, Assassin's Creed and Far Cry*), and developed franchised games with strong consumer impact (e.g. Peter Jackson's *King Kong* or James Cameron's *Avatar*). Like many creative organizations with multiple projects, the studio fits the description of a project-led organization (Hobday, 2000), with a portfolio of approximately 15–20 projects in parallel. The projects are managed through a "classic stage-gating process," which implies some very strong sets of creation, conception and production routines very well assimilated by project team members. Each project is independent and the project manager literally acts as a semi-autonomous entrepreneur, under local control of the studio's president and under the *ad hoc* and remote control of the marketing and creative department from the headquarters in Paris.

Figure 1: Assassins Creed, one of the biggest blockbusters of the Ubisoft Montréal studio.

Source: Ubisoft.

Nevertheless, the exceptional performance regarding innovation of the studio of Ubisoft Montréal during these last 20 years cannot be solely understood by the strict respect for these classical rules of management. Beyond the respect of these formal frames, the dynamics of innovation of the studio is also based on the continuous mobilization of those active units of creativity of Ubisoft Montréal: the "communities of innovation". Under this term we regroup communities of practice, as well as other forms of informal groups such as virtual communities: users' communities, epistemic communities (communities aiming at the creation of new knowledge), etc. The remarkable point in the conduct of the studio of Ubisoft Montréal is that these communities are not considered as marginal activities at the periphery of the central functioning of the organization, but quite at the opposite, they stand as the core engine of the innovative dynamics of the organization.

The secret of the success of the studio holds in the capacity of the top management of the organization to orchestrate these various communities in order to innovate on a continuous basis. The top management succeeded in coupling, in a subtle and effective way, the informal functioning of the communities with the formal and strict rules of project management that are evoked above. In this section, we underline the contribution of some of these communities: the internal communities to the company (or "communities of specialists"), and the external communities (users' communities and different professionals' communities).

I. The (Internal) Communities of Specialists

The principal sources of creativity at Ubisoft Montréal's studio rely on the functioning of "communities of specialists" (*Communautés de métier*, in French): script writers, game-designers, graphic artists in 2D and 3D, sound designers, software programmers, testers, etc. Each of these communities is composed of young professionals who are bound by emerging and weakly formalized bodies of knowledge. Members of a given community share daily information, knowledge and tricks about their work in and outside the formal framework of projects. They work in the same building, have lunch and go out together, or they just chat online with peers in search of advice or technical solutions. Within a given community knowledge is continuously exchanged and challenged, and can circulate through the existence of a local language understandable by the members only who respect the social norms of their community that drive their behaviour and beliefs (Cohendet and Simon, 2007).

These communities partly function in organized formal project frameworks, but also interact within and across their boundaries with no prescription of any rigid authority. Members of these communities are sources of specialized creative ideas, repositories of accumulated knowledge, and cooperative frameworks within which new practices and routines emerge. In each community, members communicate regularly with each other about their practice through informal cognitive spaces with more or less open boundaries, where people would meet and trade knowledge in a not-so-organized fashion, with no prescription of any rigid authority. These workspaces are not fully monitored through the formal corporate process. They are not necessarily aligned with corporate goals and strategy. They are also somewhat disconnected from the daily pressure of producing an efficient output designed for a specific market purpose. These informal socio-cognitive spaces offer areas where people can socialize, wander, meditate, confront ideas, build daring assumptions and validate new creative forms. As a result, most of the communities of specialists at the studio have a dual dimension in the way they process knowledge, aiming both at exploration and exploitation. The respective intensity of exploration and exploitation certainly varies from one community to the other (for instance, the community of game designers probably has the most weight on exploration). However, the coexistence of many diverse communities having both an exploration and exploitation dimension is in our view one of the distinctive characteristics of cultural industries and one of the reasons why these types of organizations finally succeed in matching creativity and efficiency.

Progressively, the top management of the studio acknowledged the creative potential of these internal communities and decided to boost the development of ideas and creative applications as well as the ability of these employees to solve problems. Through multiple initiatives, including the establishment of forums to connect these employees to each other, or the establishment of a team called "Guerrilla Management" (Tremblay, 2005) responsible for detecting the holders of original ideas in the organization and putting them in contact with each other, Ubisoft has gradually managed to bring out its dynamic communities of specialists in a bottom-up process. Progressively, they used the forums, and began to gather (notably on the firm's terrace), to share their expertise with other specialists internal to the firm and reciprocally help each other to quickly solve problems.

To accompany this movement, the hierarchy gave birth to new practices like the Cool Tuesdays, for every employee to share their experiences, war stories, projects and ideas; or the Hot Fridays, for producers of a game to give open feedback on the progress of their projects (inverser). These initiatives were ways to make people interact, exchange and to enact informal communities inside the organization.

This internal development has allowed for the creation of a multitude of ideas to strengthen the internal dynamics and contributed to the generative creativity of the studio. But as the company realized that a lot of their employees where working on their own projects at home, or outside the firm, the sense of urgency to change the formal ways of doing increased. These informal creations sometimes lead to external spin-offs by former employees, and were mainly a potential creative material that was unused. In order to avoid this phenomenon, Ubisoft decided to try something new and let the studio open weekends, as a way for its employees to brainstorm together and be able to develop their personal projects with the tools of the organization in total autonomy. As a result, Ubisoft is currently

Figure 2: Ubisoft's roof terrace: An important part of the internal middleground of the studio.
Source: Ubisoft site.

developing its own internal laboratory, the Fun House, to capture the ideas and game concepts (from employees but also from independent external people). They could then co-produce low budget games within the firm with them, something that was not part of the initial business model of the company. With the purpose of spreading even better practices and highlighting the various internal crafts, the studio has decided to dedicate one week to in-house events, the UbiDays, where the most talented employees animate conferences, master classes and workshops. But Ubisoft quickly realized that in order to stimulate these events further, they needed to seek local and international talents, something that was already an anchor in the organizational culture.

II. The Internal Communities of Specialists Tapping and Nurturing the "Middleground" of Montréal

In order to stimulate the creativity of its employees, Ubisoft strongly supports them (financially and materially) to participate in local events and cultural activities organized by various unions, associations or clubs in the fertile soil of Montréal. In doing so, the studio encourages its employees to freely explore new creative avenues, to get inspiration, ideas, new knowledge, and to develop their critical and aesthetic look outside the formal boundaries of the firm. Thus, the organization delegates part of its capabilities to the various communities that will tap into the local milieu of the city, to sense new opportunities that will be in some cases seized by the firm, and in some cases only, bring back commercial value to the organization. Among the many events that Ubisoft sponsored, we found the Fantasia festival that gathered more than 120,000 people over 22 days of events to publicize artists and films at the crossroads of various fields, emphasizing Asian original creations. It is a sort of "conscious investment in a staggered form of creation" from the firm, which serves not only to capture new knowledge and ways of doing things, but also plays an important role in detecting talent or offers the scope for partnerships with the outside world.

The studio is multiplying initiatives to build in its local environment to enrich its innovation ecosystem, without knowing in advance what will come out of these challenges, festivals and events. The collaboration between Amon Tobin, an artist who settled in Montreal in 2005, and Ubisoft is a good example of the ways by which these activities can influence the commercial success of a game. A group of employees working on the project of Splinter Cell III, who were passionate about electronic music, detected the Brazilian artist during his concerts in the local scene. Convinced that his way of playing could change the user's sound experience in the game, they managed to persuade the director of musical creativity to meet the artist. Following the meeting, a contract to perform the soundtrack of the game was quickly concluded. It led to even further partnerships with the artist. Amon Tobin has not only participated in the commercial success of the game, but also released an album co-produced by Ubisoft, which has been a huge success. Other similar partnerships, such as the one signed with Coeur de Pirate, have emerged this way.

The company also invested a lot in its neighbourhood to support the development of informal links between artists and local artisans, and employees of the company. By revitalizing the neighbourhood through events (such as the Ateliers du Mile end), or by investing in new forms of property in order to fight gentrification and encourage the co-location of heterogeneous entities within the same building, the firm seeks to boost the creation of social ties with actors of its local ecosystem. From this entanglement with the local, a large number of partnerships with local artisans (for the construction of luxury figurines, for example), and many more artists and entities such as the National Theater School have emerged. Partnerships with actors of the Theater School succeeded in advancing methods and techniques for motion capture, but also contributed to the development of a training program to improve the visual performances of actors. In many ways these initiatives could not have been viable if the home environment had not offered a fertile ground for individuals to build informal contacts with local artists and creators, and if it had not provided the formal institutional settings supporting the development of cultural life.

In order to tap into these local platforms of interaction, Ubisoft organizes prizewinning contests (one of which is called "toomuchimagination"), and academic competitions such as Academia, to identify local talents and bring them into the gaming industry. At the same time, Ubisoft also participates in nurturing these platforms. One of the ways the firm does so is by investing in the next generation of employees, through diverse activities and formations. A recent investment of more than US$8 million over 5 years was made in a program, in collaboration with 17 educational partners, called CODEX. This program encompasses several initiatives at all levels of education to "embrace the stages of production of a video game as a learning tool", promote the video game industry to the future generations, and offer to students a program to become familiar and train for a world of technology and connectivity. This program for the future generation will affect the industry as a whole, by educating the youth in the region in the field of video games, creating day camps for young children, or developing activities with primary schools to university classes. The growing number of independent game developers in the Montreal area reveals the enthusiasm that the industry has succeeded in creating around it, through these initiatives that each one of the studios in the territory contributed to build.

So, thanks to the communities of specialists, Ubisoft has progressively developed in the region of the Greater Montreal a rich and vibrant ecosystem that is a unique source of (re)generation of ideas, of attraction and revealing of talents. New economic opportunities of a different nature from which new business models are continuously formed and tested by Ubisoft, sometimes for the company alone, most of the time through coupled initiatives with other economic actors, can also be found in the region. The core of the ecosystem is a "middleground" or a common platform facilitating different forms of creation and exchange of knowledge between diverse communities, and continuously connecting the formal entities with the informal active units of the local environment. This platform is a "common local", partly nurtured by Ubisoft, partly supported by other formal organizations (including competitors in the videogame domain), and partly orchestrated by public local authorities. Members of the communities of specialists who are employees of Ubisoft permanently communicate with the outside world, through global virtual platforms with specialists of the same focus of knowledge, sometimes even with members of competing firms who share the same interest for a given practice. They have planted deep local roots in the creative city of Montreal, a large-scale forum consisting of a myriad of creative communities, which is a fertile soil for igniting sparks of creativity. Through this constant opening to the external world and the permanent search for the best practices from outside the organization, *communities of specialists* at the studio are unique devices tapping into the external world to bring permanently useful knowledge and creative ideas to the firm. Thus, tracing the sources of creativity at Ubisoft reveals a maze of creative communities of different sizes and scopes, a "hidden architecture of creativity" which starts from the different elementary communities of specialists of the firm.

III. The Virtual Communities of Users

Another recent and fast-growing concern of Ubisoft is the increasing role played by large communities of users (in particular virtual communities of gamers). Virtual communities of consumption, such as brand communities, create value for firms in different ways: they support a product or service, promote a brand and spread loyalty to a product or firm, or act as a resource for ideas (Carlson *et al.*, 2008). Firms thus try explicitly to utilize these communities of users to create and appropriate value for themselves. Consequently, the relationship between the firms and these communities has become an important part of the industry's business model. The industry even witnessed the emergence of firms that essentially base the value of their products on the interactions between users.

There are different types of communities of users: (a) Tester communities correspond to users whose main activity is to test games at different phases of development. In the early phases of the development of a video game, the firm uses tester communities for beta testing, mainly to search for errors, bugs or misspecifications in the program. As the development of the product matures, the firm tends to employ the community as a creative complement: for instance, the testers give advice on features to be included or excluded; (b) Player communities use specific technological artefacts to enhance or fine-tune the game or produce additional content, authorizing other users to try their creation or help the community to work better based on this fine-tuning. Player communities are primarily related to a specific game, and only through their admiration for the game do they become interested in the firm and eventually in its other products; (c) Developer communities are users who have computer skills allowing them to produce programs or to record some parts of the product, and to regularly exchange their creation with others. In some cases, firms develop parts of games with the help of users or user communities. Co-development between firm and communities (Neale and Corkindale, 1998) can be found on social software-enhanced websites or games proposed by social gaming or casual gaming companies, such as gameforge, Zinga, BigFish, etc.

Communities of specialists directly interact through different routes with communities of users, generally through social software devices: Users, video-game players or *gamers* can be considered the experts in this field; as such they are an important source of knowledge, which to a great extent circulates through informal channels that lead to communities of specialists. More precisely, the relationship with gamers is dealt with from two perspectives: from the top, in a quite structured and formal administrative way, and from the bottom-up in a very diffuse, fuzzy and emergent way. In a cultural industry such as the video-game industry where managers must analyze and address existing demand while at the same time using their imagination to extend and transform the market (Lampel *et al.*, 2000: 263), this dual perspective related to the relationships with the users is a key source of success. From the top, it is a strategic issue to understand the general trends of the market. Beyond the formal sales report, an editorial committee, a short list of historical game-designers of the company working with a chief director, works in determining the present and future interests of gamers. This team would play videogames, attend international gaming events, read about the industry and generally immerse itself in pop-culture to define the content orientations of Ubisoft. Strategic decisions to launch a new project or even to create a new brand would involve this committee, some marketing experts and even the president of the company.

Knowledge from the users seems to be more integrated from the bottom-up. As with most industries born from an almost "underground" activity, the video-game industry tries to stay close to the customers by hiring "hardcore" or so-called "lead" users (Von Hippel, 2001; Thomke and von Hippel, 2002). As gamers making games for gamers, they identify themselves with the same iconic milestone games they played along the history of this emerging entertainment form. They use a common language about the world of videogames. They would most likely agree on the generic features differentiating a good game from a bad one. This common cognitive platform allows them to work collectively as designers and also to promote their passion for videogames. This approach compensates for the difficulties in establishing a real dialogue with the elusive "casual gamer". This protean gamer plays occasionally, maybe two to three times a week, and is really difficult to capture from the marketing point of view. This part of the market is largely "ignored" by the industry (GameVision, 2005). Ubisoft and the industry in general adopted a generic branding strategy based on spin-offs from the movie industry and/or sequels of pre-existing games. Although this approach seemed to work up to now, industry experts argue that it poses a threat to creativity in the video-game industry in the long range.

IV. The External Communities of Professionals

A growing tendency in videogames is the co-creation with communities of professionals: architects, historians, geographers, sociologists, urban planners, etc. are invited to participate in the conception phase of games, in order to guarantee the quality of details and atmosphere. As an example, in a game such as Assassin's Creed, the historian's major task in the project is answering questions concerning the history of London, Paris or other places. Several questions asked include the ratio of men and women in the streets, the age when children start working, the type of shoes worn by inhabitants at this period, the precise details for roofscapes and chimney stacks and even what specific hour of the day the bell of Notre Dame or St. Paul's Cathedral ring. The studio is currently interacting with different communities of professionals who contribute, with meticulous attention to details, the aesthetic and the visual representation of the game. What is remarkable is that not only these (mostly virtual communities) contribute to the co-creation phases of the games, but they also become

Figure 3: Assassin's Creed, "The syndicate": London in 1860.
Source: Ubisoft site.

users of the game and promoters of the diffusion of the game to their social relationships. Moreover, when considering the community of historians, for instance, they realize that the game could thus be used as a historical teaching aid. As narrative medium of the digital age, games can be used in classrooms to decipher, unpack, augment and supplement the classical academic papers and textbooks. For the video-game industry, such new perspective with the community of historians could be seriously considered as a new potential business model.

V. Coping with the Communities: The Hierarchy as an Orchestrator

The development of this contribution brought forward a fundamental issue in terms of management: to a large extent, the video-game industry is facing a fast-growing situation where a significant part of the value is now created by cognitive resources (communities) that are not directly controlled by the hierarchical structures of the firm. Such a situation corresponds to the vision of the firm once suggested by Brown and Duguid (2001). They argued that the firm could be viewed "as a collective of communities, not simply of individuals, in which enacting experiments are legitimate, and separate community perspectives can be amplified by interchanges among communities. Out of this friction of competing ideas can emerge the sort of improvizational sparks necessary for igniting organizational innovation". The key issue is the integration forces implemented by the managers of the firm in order to bind the creative units together for achieving effective production, timely delivery and, ultimately, commercial successes. It appears that the nature of the relationships and ties that bind the scattered communities together is generally not a unique platform (such as a given production line or a given modular structure). These communities exchange knowledge through different cognitive platforms which are shaped or enacted by the hierarchy and which have some plasticity and flexibility to take different forms of coordination and may reconfigure through time. The growing emphasis on the role of user communities, including crowdsourcing aspects, (Noveck, 2009) leads to revisit the firm as a constellation of communities, where significant domains of knowledge production and accumulation are delegated by the firm to diverse communities, in particular communities of users. In the context of video games, some users are now able to develop and extend products or technologies, and the distinction between user and producer, or user and doer may disappear, especially with the development of the Internet.

In such a context, the main challenge for the top management is to "harness" the communities. According to Dahlander and Magnusson (2008), harnessing a community means: "(1) accessing communities to extend the resource base; (2) aligning the firm's strategy with that of the community; and (3) assimilating the work developed within communities in order to integrate and share results". From the managers' point of view, this ability to harness is the key to the success of the alchemy of combining heterogeneous communities to reach a creative collective video-game product. To go further in this direction of research, we will then develop the idea that the integration forces put forward by the firm are not just for harnessing creative units: they also generate *creative slacks* for further expansion of creativity.

This notion of *creative slack* purposefully refers to the notion of *organizational slack* proposed by Penrose (1959) who suggested that organizations always have some stock of unused or underused resources (e.g. knowledge, relationships, reputation, managerial talents, physical assets, etc.) or "organizational slack", that inevitably accumulate in the course of developing, producing and marketing any given product or service. In her view, these unexploited or underexploited productive resources are the primary factors determining both the extent and direction of firm growth. At Ubisoft, our view is that the organizational slack is essentially a creative one, which plays the role of an important reservoir of opportunities to gain innovative knowledge for the organization, and guides to a large extent the growth of the organization. In line with Penrose's vision, the firm which has accumulated a creative slack is better prepared than any other organization to benefit from the creative potential of the slack. The creative slack is shaped by the culture of the firm and is essentially understandable through the jargon of the organization. Because of these idiosyncrasies, it is much cheaper to valorize the slack *within* the firm which holds it than through any other organization (including through any isolated communities). Some may argue that the creative slack appears as a *cushion of redundancy,* which is costly to maintain. We consider that the specific conditions of formation of the creative slack at Ubisoft, which rely on the functioning of autonomous communities that naturally take charge at negligible costs of the production and conservation of knowledge in their domain of specialization, is a guarantee of the efficiency of maintaining the creative slack at low costs. The remarkable point is that the potential of the slack is diffused in the diverse communities of specialists of the firm that have memorized (thanks to the knowledge brought by their members) parts of the learning during projects. Thus, creative slack has an ambivalent characteristic: it is a specific advantage of the firm, which is the only entity able to take benefit of it, but at the same time it is held, nurtured and maintained at rather low cost by the diverse communities of the organization, sometimes even without an explicit awareness of the managers. This creative slack may also be positively influenced by the existence of multiple projects, where each project acts as a source of knowledge creation and literally feeds the members of every knowing community involved in the project, indirectly increasing the creative potential of all communities and of the firm.

To sum up, the above developments have illustrated the complex maze of creativity on which the sources of new knowledge and ideas of the studio rely. Starting from the in-house communities of specialists, we have strolled along the knowledge platforms that guarantee through non-predictable encounters, meetings, conflicts, or events, the existence of a considerable potential of creativity. Each of the communities that have been investigated can be considered as a specialized source of creativity for the company. The potential of creativity of any community is exposed to risks of inter-communal conflicts, autism or parochial partitioning; and secondly, different communities are not necessarily all homogeneous, convergent or aligned toward a common objective. Since each community is specialized in a given field of knowledge, the integration of diverse bodies of knowledge in a common thread in order to realize a creative output requires a concordance of interests and objectives between communities which is far from being spontaneous: The conflicts between communities are frequent, especially between communities working on the same epistemic domain (e.g. different communities of gamers). There is thus at a collective level a need for integration to mitigate these risks and to guarantee the systematic concordance of interests and objectives of the different communities on the spot. Management activities endeavour to establish a shared context where people could meet, debate, confront and even challenge themselves almost instantly. This occurs through physical organizations (as simple as same building, open space work areas or setting the desks in circular arrangements, see Figs. 1–4), organizational information and communication processes and also through the project manager's own open attitude towards people and

Figure 4: Ubisoft studio in Montréal.

Source: Ubisoft site.

knowledge. Organizational information and communication processes, and also through the project manager's own attitude towards people and knowledge.

That is the reason why, to avoid any lack of creativity and narrow vision due to an excessive specialization and any risk of parochialism and conflicts, the integration of different bodies of specialized knowledge within a given organization is required. The role of integration (and the justification for the existence of a firm) in such a creative domain is to assure the coexistence of diverse knowledge structures for eliciting at the collective level the sort of learning and problem solving that yields creativity. Assuming a sufficient level of knowledge overlap to ensure effective communication, interactions across communities who each possess diverse and different knowledge structures will augment the organization's capacity for making novel linkages and associations — thus fostering creativity — beyond what any community can achieve, Now, the question is to understand how the firm implements the processes of integration that harness the special sources of creativity.

References

Brown, J. S. and Duguid, P. (1991). Organizational learning and communities-of-practice: Toward a unified view of working, learning, and innovation. *Organization Science*, *2*(1), 40–57.

Carlson, B. D., Suter, T. A., and Brown, T. J. (2008). Social versus psychological brand community: The role of psychological sense of brand community. *Journal of Business Research*, *61*(4), 284–291.

Cohendet, P. and Simon, L. (2007). Playing across the playground: Paradoxes of knowledge creation in the videogame firm. *Journal of Organizational Behavior: The International Journal of Industrial, Occupational and Organizational Psychology and Behavior*, 28(5), 587–605.

Dahlander, L. and Magnusson, M. (2008). How do firms make use of open source communities? *Long Range Planning*, *41*(6), 629–649.

Hippel, E. V. (2001). User toolkits for innovation. *Journal of Product Innovation Management: An International Publication of the Product Development & Management Association*, *18*(4), 247–257.

Hobday, M. (2000). "The project-based organisation: An ideal form for managing complex products and systems?" *Research Policy*, *29*(7–8), 871–893.

Lampel, J., Lant, T., and Shamsie, J. (2000). Balancing act: Learning from organizing practices in cultural industries. *Organization Science*, *11*(3), 263–269.

Neale, M. R. and Corkindale, D. R. (1998). Co-developing products: Involving customers earlier and more deeply. *Long Range Planning*, *31*(3), 418–425.

Noveck, B. S. (2009). *Wiki Government: How Technology can Make Government Better, Democracy Stronger, and Citizens More Powerful*, Brookings Institution Press, Washington, DC.

Penrose, E. (1959). The theory of the growth of the firm. *John Wiley& Sons, New York.*

Thomke, S. and Hippel, E. V. (2002). Customers as innovators: A new way to create value. *Harvard Business Review*, *80*(4), 74–85.

Tremblay, D.-G. (2005). Les communautés de pratique: Quels sont les facteurs de succès. *Revue internationale sur le travail et la société*, *3*(2), 692–722.

© 2021 World Scientific Publishing Company
https://doi.org/10.1142/9789811234286_0003

Chapter 3

Salomon Breathes New Life into its Innovation Approach Through Sports Communities

Benoit Sarazin and Jean-Yves Couput

Salomon is a company that has succeeded in gaining a sustainable competitive advantage from its relationship with the user community of its products. It is a delicate relationship that has grown gradually and has required profound adjustments. This chapter describes how a user community is born and evolves. It shows the benefits that a company like Salomon gains from its relationship with this community, both in designing its products and in distributing them. It describes how Salomon structured its relationship with the community. And it brings out best practices as well as traps that should be avoided.

I. The History of Salomon and the Communities

1. *Right from the word go, Salomon has always had strong links with the community of sports enthusiasts*

The Salomon brand began as a family business in the Haute-Savoie region of Eastern France, bordering Switzerland and Italy. Georges Salomon transformed the workshop his father had created, which fabricated the edges of ski blades, into an internationally renowned business specializing in accessories for mountain sports. Throughout his career, he has built his success on two elements: the quality of the products, and an attentive, sustained relationship with the very best athletes. As such, the brand regularly launches innovative products that radically change athletes' performance levels, and consequently it has profoundly transformed the sport. For instance, Salomon invented a new generation of binding that fastens the skis to the boots, an innovation that was created in collaboration with the skiing champion, Emile Allais. The company's culture of being open to working with athletes naturally places Salomon at the heart of the community of sports people. Indeed, the vast majority of its directors and employees are, themselves, great sports enthusiasts who do not hesitate to test the prototypes they are developing.

When faced with a problem, the Salomon team instinctively takes the type of approach one might expect of a community of innovation, without necessarily even knowing they are doing so. This is what happened in the 1990s, when Salomon came up against a major challenge: the market they were selling to stopped expanding. The brand was essentially operating in the winter sports market — a sector whose potential for growth was weak, and whose activity was greatly reduced in years when there was little snowfall. As a result, the company looked to diversify by expanding into outdoor sports, practiced throughout the year. When confronted with a situation like this, most companies would limit themselves to finding a niche within one particular segment of the existing outdoor sports market, for example, specializing in hiking or trekking shoes. This was not the case for Salomon. The Salomon team adopted the attitude of an innovative community, constantly seeking to enhance the surrounding environment. The company chose to help new practices emerge. The mountain in the summer season was considered by most people as a space for gentle leisure pursuits, for walking and exploring the scenery, in contrast to its use for more fast-paced sports in the winter. Salomon, however, had a different intuition: they thought that the mountain could be more than a playground — they believed it could become a sports field.

This is the approach followed by Jean-Yves Couput, head of "sport marketing", the Salomon department that provides technical support to top athletes. In 1998, he developed a new sporting practice known in French as "Raid Aventure", and which exists in English-speaking countries by the name of Adventure Racing, or Expedition Racing. He was inspired by the "Raid Gauloise", created by Gérard Fusil in 1989. "Raid Gauloise" is a kind of endurance test featuring diverse physical challenges, where teams of athletes must cross the jungle over the course of a week by mountain bike, kayak and on foot. Couput conceived "Raid Aventure" along similar lines in terms of its physical challenges, though it was more athletic and involved a shorter course, lasting just two days. In 1998, Salomon launched the first "Raid Aventure" competition. In 1999, Salomon organized 7 international events and asked its subsidiaries in the main countries to sponsor teams of athletes who participated. Couput set up press tours where journalists could have an inside view of the event and experience it by running on some portions of the race. The celebrity status the runners gained inspired the most entrepreneurial members of the community to become event organizers and create their own similar events on a national scale. As a result the sport developed, and Salomon no longer needed to sponsor the events. In 2012, there were more than 500 "Raid Aventure" events organized across France.

As time went on, and aware of the logistical difficulties involved in transporting the bikes and kayaks needed for the "Raid Aventure" events, Salomon progressively put more emphasis on trail running in their adventure race format. The brand eventually shifted the nature of the competitions towards trail running — a sport purely about running on trails in forests, mountains and deserts. Trail running has become an attractive practice for those who want to be both physically active and to spend time in nature. Trail running events have gained even more popularity than "Raid Aventure", with more than 10 million participants worldwide, and 3,000 competitive races in 2017.

2. *In 2008, Salomon made a decisive shift: To focus its marketing on the community*

In 2008, Salomon found itself in something of a paradox. The company had gained success in the emerging market of trail running shoes: it had caught the eye of runners, and Salomon's range of shoes had become the point of reference in the trail running world. However, the company was at risk of losing out to bigger brands, who had far greater marketing budgets.

Indeed, Salomon had secured its success on a limited budget, which was reserved uniquely for athlete sponsorship. Since trail running was still a relatively small market, and since running was a minor sector of Salomon's activities, the company chose to dedicate its communications budget to winter sports, which at that time represented its core business.

With little to no communications budget, Salomon's trail running business risked being knocked out of the water by bigger brands who had far greater marketing capacities. This threat became a reality: attracted by the growth of trail running, all the big running footwear companies launched their own models of trail running shoe. They bought in famous athletes and sponsored competitions, outbidding each other's investments. The stakes were clear: if Salomon continued with traditional marketing strategies, it was at risk of being squeezed out of the market by its rivals. Salomon had no choice but to transform its marketing approach.

Salomon's management team then called on a consultant, Benoit Sarazin, to help them decide on which path to go down. Benoit is a specialist in disruptive innovation. He acts as a catalyst for his clients. He helps them think differently, identify opportunities that are accessible to them, and build action plans to capitalize on these opportunities. In 2008, the notion of community marketing — that is, the form of marketing that capitalizes on fan bases to promote products — was still in its infancy. However, this was the reason for Salomon's success in Trail Running: the practitioners of the sport formed a close-knit and dynamic community of enthusiasts who recommended Salomon's products to their entourage. Benoit helped Salomon's team become aware of the major role that the community played in the success of the company's products. He showed that Salomon paid little attention to the specific needs of this community and was risking letting a competitor become more attuned to the community's needs and ultimately become the latter's favored brand. He also pointed out that the traditional marketing efforts that the team were considering were not very effective: advertisements in specialist magazines and event sponsorships had little weight when faced with larger competitors. He recommended that the team focus their marketing efforts on the community and structure the relationship between Salomon and the community. This new direction allowed Salomon to gain an advantage over its more powerful competitors in spite of a more limited budget. With the new orientation, Salomon has strengthened its dominance in the trail market, enjoying a more than 50% market share in 2011 and, despite competition, has maintained a leading position in this market.

In 2011, Salomon took up the Barefoot Running challenge with the help of its community.

In 2011, even though it led the market for trail running shoes, Salomon was confronted with a new trend which called into question its strategy. "Barefoot running", a technique inspired by a return to what is more natural for the body, began to catch the eye of runners. As the name suggests, barefoot running involves running barefoot, or with minimalist shoes. This trend developed from a basic principle: that the human body was designed for running, thus does not need modern, shock-absorbing shoes. It is entirely possible to run barefoot, so long as you alter your gait accordingly: rather than striking down on the heel with each stride, you run instead on the ball of the foot, known as "forefoot running". The concept was popularized when the journalist Christopher McDougall wrote a book entitled *Born to Run* in which he showed that the best endurance runners in the world were Native Americans running barefoot. The brand Vibram brought this concept onto the market by inventing the "FiveFingers", a minimalist shoe that slips onto the foot like a glove and simulates the sensation of running barefoot. Vibram maintained that this shoe minimized the risk of injury, and they became an overnight success.

For Salomon, this trend was bad news. It ran counter to the design principles behind Salomon's trail shoes, created with "protection first" in mind: their aim is to shield the runner's feet against obstacles such as rocky terrain, tree roots, or thorns. The marketing team at Salomon really started to scratch their heads. What on earth should they do? Should they ignore the "barefoot" trend? If so, they risked their clients turning away from the brand, instead gravitating towards minimalist shoes in the hope of getting "back to nature". Should they throw themselves into the trend for minimalism, at the risk of contradicting the principles of solidity and comfort which had earned them their reputation?

The answer to this thorny question was eventually found through a collaboration between a "lead-user" (Hippel, 1986) and three members of the Salomon team: an engineer, a designer and a podiatrist. The lead-user was Kilian Jornet, one of the top trail running athletes in the world, and a figure whose charisma inspires other members of the community. Salomon has known him since he was 14 years old and, ever since, has supported him in following his passion for running in the mountains. The brand helps him develop his online presence, and provides him with made-to-measure shoes. Patrick Leick, an engineer, is in charge of creating these bespoke shoes for Kilian — tailormade adaptations of existing models. The close relationship between Kilian and Salomon's team is also illustrated in Figure 2, where Greg, the director of Sports marketing, collects on-the-spot impressions during a race, and in Figure 4, where Kilian gives his feedback on the performance of his shoes after the race. During the

Figure 1: Kilian Jornet in the Western States competition in 2010, an experience that will inspire a new shoe, the "Sense".

Source: Salomon.

Figure 2: Greg (on the right), the director of Salomon Sports Marketing, collects Kilian Jornet's on-the-spot impressions at 70 km during the Western States competition.

Source: Salomon.

Western States run in California — a 100-mile race along the precipitous route that the Gold Rush miners used to cross the mountains — Kilian noticed the limitations of existing shoes.

After a first failed attempt, he was determined to come back the following year to win the race, and he asked Patrick to create a shoe that could offer even higher performance capacities. He wanted to reduce the weight of his shoes to the bare minimum and increase his ability to feel the ground beneath his feet, while still maintaining the same level of comfort and protection on rock gardens and technical trails.

Kilian's requirements could not be met by simply adapting existing products. With the help of Benjamin (the designer) and Abdel (the podiatrist who is shown in Figure 3 while taking Kilian's footprint to create a custom made shoe), Patrick designed a completely new concept from scratch. In the space of 10 months, and after Kilian had tried out 40 different prototypes, the shoe was finally ready. As shown in Figure 1, Kilian used it and in June 2011 won the Western States. Since Kilian was a forefroot striker, like all enthusiasts of barefoot running, this product met the requirements of minimalist shoes. Yet it also went further, protecting the

Figure 3: Abdel, podiatrist, takes Kilian Jornet's foot print in order to create a custom-made shoe.

Source: Salomon.

foot against stones and allowing him to run fast downhill — features that the "Fivefingers" could not provide. With such shoes, Killian was able to run comfortably along 160 km of mountain trails.

The shoe thus built became the source of a successful product. Initially, this initiative was not a part of Salomon's development plan. It was only conceived in response to a specific need expressed by a high-performance athlete. However, once the marketing team at Salomon realized the strengths of the product created for Kilian, they put it on the market under the name "Sense". This innovation gave Salomon several advantages. First, it was a high-quality response to the trend for minimalism, without moving away from the principles of comfort and protection which had secured Salomon's success in the world of trail running. Furthermore, this new shoe helped prevent the risk of injury for those inexperienced runners taking the plunge into barefoot running with little concern for the sport's potential risks. Although running on the forefoot does have its benefits, it nonetheless puts a great deal of stress on the tendons and metatarsal bones. When inexperienced runners suddenly switch from a traditional heel-striking gait to running on their forefoot, they run the risk of tearing their calf muscle, rupturing their Achilles tendon, or fracturing one of the metatarsal bones. Salomon is acutely aware of these risks; they ensure that the "Sense" remains a high-performance product, specifically intended for experienced runners, while guiding less experienced customers towards more traditional trail shoes which do not present this type of risk.

Salomon's push to shift the centre of interest from barefoot running to natural running ended up giving them a strategic advantage over Vibram. Vibram adopted an aggressive marketing approach in seeking to convert as many runners as possible to barefoot running. Unfortunately, many of these converts sustained serious injuries, and in 2014 rumours of the dangers of barefoot running spread like wildfire among those practicing the sport. Sales of minimalist shoes tumbled. Pursued by the courts for false advertising, in May 2014 Vibram had to pay out US$3.75 million in compensation to reimburse dissatisfied customers.

This story illustrates the power of the collaboration between Salomon and the trail running community. Indeed, Kilian was not an ordinary lead-user. He was one of the leaders of the Trail Running community: he inspired practitioners with both his sporting achievements and his humble attitude towards the mountains. As a member of the community, Kilian had access to the community's "creative slack". The creative slack is the

Figure 4: Kilian Jornet gives his feedback on the performance of his shoe after the race.
Source: Salomon.

pool of creative ideas that members of a given community (of enthusiastic practitioners of a sport, for example) generate spontaneously thanks to the inspiration they get from following their passion. These ideas usually remain untapped, or if they are tapped, they rarely go further than the prototype. In the same way, Patrick was not an engineer like any other. He was part of the community of technology enthusiasts active in the field of sports shoes. He too had access to the creative slack of his community and was well aware of both the possibilities of technology and the tests being carried out by colleagues or competitors. Benjamin, the designer, and Abdel, the podiatrist, were both passionate about sport and had access to the creative pool of the designer and podiatrist communities. When Salomon arranged for these four individuals to work together, the company encouraged the combination of ideas from the creative reservoirs of the four distinct communities. It gave these ideas the means to become realized in the shape of an innovation. On the other hand, since Kilian is a leader who is admired by community members, all the runners wanted the same shoe as him. As soon as the product was available, the most high-performing runners immediately adopted it. By tapping into the creative reservoir of communities, Salomon found its response to the trend of barefoot running.

II. The Benefits that Communities Bring to Salomon

The benefits that Salomon derives from its relationship with its communities can be found in three areas:

- Collecting inspiration from the community to imagine new innovative products. In particular, the relationship between Salomon and Kilian Jornet, one of the leaders of the trail running community, is particularly fruitful. Kilian constantly pushes the boundaries of sport and encourages Salomon to make major innovations such as the "Sense" shoe.
- Creating and fine-tuning innovations through involving community users. In the Sense example, the shoe was developed through the creation of many prototypes that Kilian tested to validate the shoe's design.
- Distributing the products on the market. The community's influence on the market is explained by the communicative passion of its members. The community's core is composed of enthusiasts who share their passion for the sport; these individuals influence less devoted members and offer advice to those new to the sport. This core of enthusiasts recommends Salomon products because they meet their requirements, and because the brand contributed to the development of the sport during the very early stages. They act as precursors who diffuse their preferences to others.
- Getting retailers on board. By nature, retailers are risk-averse; looking for guaranteed returns, they are reluctant to begin selling products for new sporting activities, and prefer the security offered by established sports. A brand like Salomon had little clout to convince retailers to dedicate a space in their shop to trail running, given that this sport had not been around long. But when sporting enthusiasts, who were also regular clients, began to spontaneously request trail running shoes, they weighed up the new sales potential and entered into business with Salomon.

III. Community Dynamics

1. *How does a community come into being?*

A community is formed when a group of individuals sharing the same passion suggest creating a movement in complete contrast to any that currently exist. They take on a leadership role, as founders of the community. An essential step in the creation of the community is when the founders come up with a group manifesto. This manifesto can be written down, but it might also remain tacit and undocumented.

The point of the manifesto is to provide a community identity, and to state its aims and values. It defines exactly what the community hopes to create, and which does not currently exist. It also explicitly clarifies what the community rejects. Crucially, it acts as a kind of code of belief to convince others to join the group.

An example of a manifesto is the one that Kilian created when he started his project, "Summits of my life". Although this project is centred on a personal goal (to beat the records for ascending and descending the highest peaks in the world), Kilian's ambition stretches much further than this. He calls on a large community of supporters and partners whom he makes his associates. He offers them a new, purist, minimalist way to look at mountain pursuits. On his website, Kilian lists the values of the community (all available at http://summitsofmylife.com/fr#/valores):

1. *No one told us what we were. No one told us we should go. No one told us that it would be easy. Someone once said that we are our dreams. If we don't dream we are no longer alive.*
2. *We walk in the footsteps of instinct leading us into the unknown.*
3. *We don't look at the obstacles we've overcome, but at those we've got ahead of us.*
4. *It's not about being faster, stronger or bigger. It's about being ourselves.*
5. *We're not runners, alpinists or skiers...we're not only sportspeople... we're people.*
6. *We can't be sure we'll find it, but we're going in search of happiness.*
7. *With simplicity.*
8. *In silence.*
9. *Responsibly.*
10. *What are we after? Might it be life?*

2. *The community, an unstable entity that grows and splits itself*

Once the community has been created, the passion of the founders and of their followers moves them to evangelize and recruit new members. At the same time, though, the community is selective. It welcomes members who share the vision set out in its manifesto, yet rejects others. New members, even though they adhere to the common vision, add their own, unique preferences. This rich diversity of perspectives forces the community to evolve and to move beyond the ideas of the initial manifesto. And so we reach a paradox. It often occurs that new members put forward new ways of thinking that the founding members reject as incompatible with their original vision. This inevitably leads to a scission, and a new community emerges with new values and beliefs.

Let's take the example of the community of mountain bikers. In the beginning, mountain biking was a sport invented among the hills south of San Franciso by a few individuals with a passion. For fun, they took their bikes up into the mountains. The sole aim was to enjoy themselves and have a good time riding downhill. Gradually, the practice spread and road cyclists and cyclo-cross enthusiasts joined them. They brought their experience of endurance races with them, and the practice shifted towards competitive events, with races where cyclists sprint from one point to another as quickly as possible. In time, film and TV became interested and imposed their own constraints. Races from one point to another were tricky to film, so the routes turned into circuits with the same start and finish point, where competitors were required to complete several laps. These changes frustrated the original founders of mountain biking, for whom the pursuit was about discovering new spaces to ride. The community split into several branches which in turn evolved into independent communities: those who wish to explore new spaces with total freedom, and those who want to climb to the top of the medals podium. As time went on, these groups divided again according to different practices: long-distance mountain biking, short distance, trekking, cross-country exploring, or as a simple leisure pursuit.

IV. The Community Began Local, but Became Global

Communities have always existed. Men have always sought to share their passions with others. It is this phenomenon that gave birth to the community. However, until social media arrived on the scene, these communities were limited to the relationships that individuals could build with one another via real-life encounters. They could be formed through friendship groups, through club meets and through associations that gathered around a common interest. In all cases, the relationships remained local — restricted by the natural barriers of geographical proximity. With the advent of online social media, these communities became global. Exchanges between members were no longer limited by distance, but rather stretched right across the world. For instance, Trail Running's best athletes are known worldwide.

V. Best Practices to be Learnt from Salomon's Experience

Salomon's experience with user communities provides lessons that can be applied in other contexts, namely, on structuring the relationship with the community and on how to adopt an attitude of sharing and openness.

VI. Structuring the Relationship Between Upperground, Middleground and Underground

The relationship between sports companies and the sporting community clearly illustrates the virtuous relationship between the "upperground", "middleground" and "underground". The underground is composed of members of the community who share the same passion. In the case of sports, these are the sports enthusiasts themselves. They include elite athletes as well as those who practice the sport seriously or just occasionally. The upperground is the group of businesses and institutions that are involved in the sport. They include, for example, companies who make sports equipment, the chain stores that sell them, and the administrative authorities who manage public sports spaces of all kinds, from leisure centres, to stadiums, to forest trails. The upperground and underground are essential to one another. If the manufacturers of sports equipment did not constantly innovate to improve their products, sportspeople would quickly run out of ways to develop. If local authorities did not take on the management of sports pitches and leisure centres, it would be difficult to carry out these pursuits. And reciprocally, of course, without the sporting community, to whom would manufacturers sell their products? Who would use the sports centres, put in place by council authorities?

And yet, the relationship between the upperground and underground is counter-intuitive, for the reason that their values are diametrically opposed. The upperground is structured according to authority; whether we consider businesses or governmental bodies, the decisions of those further up the management chain define the operations of the upperground. At the opposite pole, the community of participants is non-hierarchical: nobody is the manager. Even if certain members are recognized as "leaders", they have no authority over other members because everyone is equal. The key value is honesty between community members. How does one member come to understand that he or she can rely on another? By the latter contributing to the interests of the community in a selfless way. We can see why the underground might be wary of the upperground. Members of the community view businesses and government authorities as groups who pursue their own interests: businesses are looking to maximize their profit, while officials are hoping to be re-elected. By the same token, the upperground is loath to rely on communities. There is no manager to do a deal with; if a member of the community commits to something, there is no lasting guarantee, for tomorrow other members might suddenly call it all into question.

The relationship between the upperground and underground is made possible by the middleground. The middleground is all those spaces and activities whose primary objective is to facilitate the functioning of the community. Research shows that the middleground's activities are centered on four axes, which are non-exclusive and mutually support one another:

- **Events:** In sports, events include competitions, festivals and professional fairs, training sessions and collective outings guided by a group leader. These are particularly important moments for community members: these are the contexts in which they have the opportunity to practice their sport with others and to talk with them. They can track and compare the results of their training; they can discuss issues they are having, and ask other enthusiasts about solutions they may have found. Together, they are able to share an experience that will remain in their memories.

- **Projects:** A good example of a community project is the collaboration between Kilian Jornet and the athlete support team at Salomon on the design of a new shoe, the "Sense". This shoe helped Kilian become the first European runner to win the Western States trail running competition, held in the US.
- **Places:** In the case of sports, these are the locations where athletes meet informally and practice their sport together. They are also social media where sport practitioners share their experiences.
- Spaces, i.e. cognitive reflections that propose to explore new concepts or new paradigms.

An example of a new "space" is Kilian Jornet's calling into question the definition of his sport. He innovated by breaking down the barriers separating trail running and Alpine mountaineering (Alpinisme in French), creating a practice he calls "AlpinRunning". These activities are traditionally poles apart, both in theory and in practice, and their practitioners form separate communities. Trail running is all about completing a mountainous route in the fastest time possible, and ascending steep climbs without ever actually reaching the summit. For example, the Mont Blanc Ultra Trail (UTMB) — the king of trail running competitions — has a route that forms a circuit around the Mont Blanc mountain range. The runners have to make it across 168 km and climb 9,600 m along mountain passes and precipitous ridges, but they never approach the summit of Mont Blanc itself. Alpine mountaineering, on the other hand, is about ascending a mountain as elegantly as possible, taking the most varied and complicated routes until you reach its summit — the duration of the race, however, is not a factor. In short, Trail Running favors detours where mountaineering searches for the purest of lines, i.e. the straight line. While still very young, Kilian was already an athlete who excelled in several sports. This is a rare thing among champions, who are often specialists in just one discipline. By 20 years old, he had already won the most prestigious competitions in both ski touring and trail running. However, his role far surpasses that of a traditional champion who operates within the bounds of the preestablished rules and respects the barriers erected between sports. With AlpinRunning, he suggested combining trail running and Alpine mountaineering to create a sport where participants must run from the bottom of the valley to the summit of the mountain — and back! — in the fastest time possible. Kilian set himself the goal of beating the records for the fastest ascent and descent of the seven highest mountain summits in all continents of the world. He beat the record for ascending Mont Blanc from the starting point of Chamonix (in other words, 3,800 m of pure ascent), taking just 4 hours and 57 minutes, and using trail running shoes fitted with metal studs in order to grip on the ice. (For comparison, when experienced sportspeople run this route with a guide, they split it over two days and leave from higher up the valley, so as to limit the length of the climb: 1,250 m from the drop-off point of the Aiguille du Midi cablecar).

VII. Contributing to the Middleground to Establish a Strong Relationship with the Community

It is through participating in the middleground that Salomon has succeeded in establishing a strong relationship with the sports community. This participation takes two complementary forms: community leadership by community managers and support for athletes. Community managers facilitate activities and contribute to a dynamic community life. They enrich the dialogue between community members with their comments on social media. They also provide support for local initiatives. For example, when a specialized store organizes an event for Trail Running practitioners, community managers coordinate various support actions: they spread the word about the event, invite well-known athletes to make the event appealing, provide products that athletes can try, offer a snack at the end of the race, and provide technicians who will help participants solve their problems. This role is essential: without such facilitation, community members disengage and the community dies. Salomon also provides personalized support to athletes. These athletes are community leaders who inspire and influence others. Salomon provides them with tailor-made equipment that contributes to their athletic performance. Salomon also supports them in their leadership role, helping them gain notoriety and helping them communicate on social media.

1. *Bringing value to the community in authentic ways*

Companies and organizations that contribute to the well-being of the community must find a way to make a profit out of their activities in order to remain financially viable. Unfortunately, the community tends to reject those who take a mercenary approach. So how can we move beyond this stalemate? The answer is for the company to bring value to the community in an authentic way, without necessarily expecting a direct gain from this action. Thus, the company must act as community members do: they share without expecting anything in return.

Take, for example, the case of Bryon Powel, who decided to leave his career as a lawyer in Washington DC to dedicate himself to his passion for trail running. He owns the blog www.iRunFar.com where, alongside a small team of experts, he posts news on trail running and ultra-marathon racing all over the world, competition results and gives his opinion on different practices and products. By creating iRunFar, he moved from being a mere member of the community to operating as an active contributor to the middleground. He funds all this by selling advertising space on his blog pages to companies offering related products. Indeed, in recent years, many brands have become aware that in order to be visible to a mass clientele, it is important to be seen as legitimate by those who are influential in their communities. But, by making a profit out of hits on his blog, doesn't he risk alienating members of the community who might see this as a mercenary action? The rapid increase in traffic to his blog would suggest quite the opposite. The reason for this is simple: the primary purpose of his blog is to serve the community by providing relevant articles which do not provide him with any direct gain, since they can be accessed for free. The adverts on his site do not contradict this purpose either, as they promote products that iRunFar deems useful to sportspeople. Bryon only profits indirectly: his blog's success with readers makes it attractive to companies seeking to promote their products.

This is the attitude that Salomon adopts when it provides personalized support to the best athletes in their domain. Salomon does not simply finance athletes in exchange for their participation in advertising campaigns. Salomon seeks to bring what is most precious to them. The brand does not hesitate to support athletes early in their careers, from their teenage years when their record is still almost non-existent. Then Salomon helps them communicate their achievements and establish their notoriety. Finally, it is one of the only brands that provides them with custom-made shoes. This is a key point given that more than 30% of athletes have feet that are different enough from the norm that standard shoes are unsuitable for them. Salomon's attitude towards athletes shows that the brand does not have a mercenary attitude, even if it is obliged to generate turnover.

Salomon's contribution to the community has a major impact on the success of its products. The more Salomon is seen to be helping the community grow, the more its positive image is reinforced among sportspeople, and the more these people — when faced with a shelf full of similar products in a shop — will be persuaded to choose Salomon over a rival brand.

2. *Being transparent in communication*

How can the company communicate to community members who distrust corporations? Indeed, the community generally believes that companies embellish the positives and hush up the negatives of their brand. Furthermore, a brand's marketing discourse can appear dehumanized due to the company being far removed from the real-life experiences of the community members it is selling to. In short, it's all seen as smoke and mirrors, designed to sell — unacceptable for a community whose core values are trust and authenticity. So, even if Salomon's employees are sportspeople closely involved in the community, the second they mention the brand they can't help come across as smooth talkers. The lesson is clear: as long as communication between the sporting community and the company takes place uniquely through the conduit of employees, members of the community will remain suspicious.

Realizing this fact, Jean-Yves Couput recruited a network of "community influencers" in the main countries where Salomon had a market presence, to help manage relationships between the brand and the community. These community managers are members of the community who have been earmarked as spokespeople for the brand. They take part in conversations on social media, giving their opinion on practices and products. Yet, since they are not Salomon employees, they have the freedom to express their personal opinion. If they think a Salomon product has a fault or that the company has taken the wrong approach, they will say so, and without mincing their words! It is this honesty that then makes them credible when they defend the brand's actions or the quality of its products.

Jean-Yves chose them among the most involved members of the community who were outside the company because they had credibility within the community. They were often athletes who were reaching the end of their sporting careers. By offering them new career perspectives at a time when they should have been giving up sports, the brand fostered sympathy capital among practitioners.

VIII. The Traps that Companies May Encounter

These principles seem simple and yet their implementation is complex. Below are the main traps that companies or organizations can encounter in the world of sports when they try to connect with communities.

1. *Trap 1 = Not respecting community values*

Members of a community have shared values that are dear to them and do not accept that these are violated. It is important that companies and organizations who wish to establish a relationship with their communities respect these values. This is not always easy. A disagreement that arose in 2015 between the French Athletics Federation (Fédération Française d'Athlétisme) and some of trail running's top athletes clearly illustrates this dilemma. Trail running competitions have long been held without the supervision of any sporting federation. Since 2008, the French Athetics Federation has been trying to offer its input in order to standardize the way competitions are organized. For the Trail Running World Championship held in Annecy, France, in May 2015, the FAF stated that the race should begin in two stages: runners belonging to national teams were to leave 2 hours before other, non-nationally ranked, competitors. Consequently, the non-ranked runners had no chance of winning the race. Now, in order to be nationally ranked, a runner must take part in competitions held in all corners of the world, planned according to a strict timetable, which requires a great deal of flexibility and significant financial means to pay for the travel. The FAF's rule provoked a boycott by certain elite athletes who argued that it was unfair. As François D'Haene, one of the top two trail runners in the world alongside Kilian Jornet, explained on his Facebook page, this regulation contradicted a core value of the trail community: that the sport is open to anyone, with all runners considered equal. For him, a competition is a celebration where all participants meet regardless of ranking or performance results: the average runner can run in the same race as the champions.

2. *Trap 2 = Exploiting the community by taking but not giving*

Let's take the example of an accessories brand that sets up a stall alongside the route of a major trail running competition, in order to display its products. This would seem to be a lucrative opportunity given that, for example, in 2015 the Mont Blanc Ultra Trail welcomed 7,500 competitors and crowds of spectators — an attractive public for the company. And yet, this action alone will not win over the community in the long term unless it is accompanied by a genuine commitment to making a positive contribution to the community. In parallel, the brand must support those who sustain the community and bring it to life: event organizers, athletes, community media, etc. This support might come in the form of financial help or by sharing the technical or human resources the company has at its disposal. Support might also be shown via the creation of web platforms where members of the community can express themselves, such as the Facebook page Salomon created which is used for posting athletes' results. This contribution is particularly valuable when the community needs help to become structured. As an example, when Bryon Powel started iRunFar.com, Salomon contributed financially to buying the equipment needed to establish a satellite connection and cover live trail events in remote mountain locations. Such contributions allow brands to establish a relationship with their communities as a first step. Later, it will be able to offer its products to community members.

3. *Trap 3 = Having a passive attitude*

Sometimes, companies might express an interest in helping organize an event. However, when it comes to planning meetings, the company's representatives take a back seat and never volunteer to wake up at 4am to put up banners and only get to bed at midnight to help take them down. Such passivity is poorly viewed by the community and will have a negative impact on the brand.

4. *Trap 4 = Having an oversimplified message that is perceived as a lie by the community*

The community expects transparent dialogue, where nothing is hidden and lies are forbidden. This is an extremely tricky thing. Even if the requirement appears fair, it is almost impossible to get right. Why should being transparent be problematic? The difficulty is twofold. First, the messages put out by brands must be simple, so as to be easily understood.

This requires them to cut down on information, which in turn makes it inaccurate. Secondly, members of the community do not always really listen to the messages they receive. They apply selective hearing and retain only the information that interests them, ignoring the rest. The answer lies in making a clear difference between an acceptable bending of the truth, versus unacceptable lies.

Take the case of selling downhill skis to the general public. When shops sell their customers "race skis", these are not the same as those used in competitive racing. This is an example of a lie that is generally accepted by the community. Why? Because many skiers like the idea of buying the same skis as those used by the champions they admire. Actually selling them such equipment would be incredibly dangerous, since these skis can only be safely used by those with a total mastery of the sport, or else injuries would occur. Thanks to their selective memory, people buying skis forget that champions have a highly developed muscle structure that helps prevent them from injuring themselves in difficult situations, and that they are surrounded by a support team who prepare their skis before each race according to the specific snow conditions. This is not the case for most skiers. This explains why skis sold as "race skis" are less rigid and have a higher tolerance for faults than those used by champions. The sales terminology is therefore an acceptable bending of the truth, because it both looks out for the safety of the customer and because the difference between the skis is known to the shop's experts.

On the other hand, Vibram's claim that its "FiveFingers" shoe reduced the risk of injury was seen to be an unacceptable lie, leading to a rejection by the community. In 2009, Vibram began promoting the "FiveFingers", the minimalist shoe that allowed runners to experience the sensation of barefoot running and to get back to a more natural running posture. It is true that "FiveFingers" reduces injuries linked to the shock of the heel striking the ground, as it forces the runner to strike on his or her forefoot. However, this type of gait can lead to other, even more serious, injuries: from muscle strain to rupturing the Achilles tendon or fracturing the metatarsal bones, for example. In spite of the videos posted by Vibram on its website explaining the importance of making a gradual transition from a traditional gait to a forefoot strike, many runners chose to ignore this advice. They switched to barefoot running too abruptly and did serious damage to themselves. In 2014, product sales tumbled, and the brand's image was tarnished.

5. *Trap 5 = Demanding exclusivity*

It is tempting for a brand to impose exclusivity to be the only one represented in an event in exchange for its support. Such a requirement is poorly perceived by the community because it runs counter to the community's values of sharing and openness. For example, when Salomon funded a special issue of the American magazine *Runner's World* to promote Trail Running, which was still emerging in the United States, the brand ensured that the products cited included those of competitors at the same level as those of Salomon. Salomon went so far as to offer its main competitor a free advertising page to promote a series of races that the competitor was sponsoring. Asking for a "Salomon only" exclusivity would have destroyed the magazine's credibility with the community.

6. *Trap 6 = Adopting an impersonal, arrogant attitude*

If a company wants to talk to the community it must do so as a human being and not as a cold, impersonal, or all-powerful entity. It has to show its strengths but also its weaknesses, which can be difficult given that companies naturally want to highlight their strong points, and the things they are sure of. This is a lesson Salomon learnt at its own expense when a customer seriously injured himself while using a pair of touring skis with a binding that did not fulfill its protective role. This skier was also a famous blogger in the United States. He vented his anger at the brand on his blog. Worried about the consequences of this post for their image, the Salomon team reacted with a stony silence, followed by formal responses drafted by their legal team. This approach only aggravated the situation and the blogger launched a boycott campaign against Salomon products. The situation was only resolved long afterwards, following a legal negotiation. Salomon discovered that in this type of situation the company needs to show its human side and admit the error of its ways. The spokespersons representing the brand should instead show empathy, and instantly engage in the conversation to understand and discuss what has happened before taking a stance, as if it were a dialogue between two people. When the company shows itself to be honest and transparent, aggression usually melts away.

IX. Conclusion

Establishing a relationship between a brand and a community of users is a sensitive undertaking. If successful, it can result in a win–win exchange: community members have the support of the brand to deepen their experience of their passion, while the company enjoys the creative momentum provided by the community, helping it create and distribute its products. But it can also have a much less positive outcome, which can go so far as the boycotting of the brand's products by users. Salomon's experience shows us that this relationship must be structured: the company gains the trust of the community by contributing to the middleground, i.e. to the activities and spaces that breathe life into the community. It also requires that the company adopt the same posture of sharing and openness as that of members of the community. In concrete terms, this means bringing value in an authentic way, for example, by sharing without expecting direct returns. It means being transparent when communicating, for example, by allowing community managers full freedom of expression. Such relationships are fragile and there are many pitfalls, such as failing to respect community values, which can ruin the relationship at any time. It is a learning experience that managers must master if they are to open the door towards a new era of innovation.

References

Cohendet, P., Grandadam, D., and Simon, L. (2020). The anatomy of the creative city. *Industry and Innovation*, 17: 91–111.

Cohendet, P. and Simon, L. (2015). Introduction to the special issue on creativity in innovation. *Technology Innovation Management Review*, 5(7): 5–13.

Von Hippel, E. (1986). Lead users: A source of novel product concepts. *Management Science*, 32(7), 791–805.

© 2021 World Scientific Publishing Company
https://doi.org/10.1142/9789811234286_0004

Chapter 4

Schneider Electric is Steering its Communities of Practice with the Communities@Work Program

Louis-Pierre Guillaume, Catherine Thiesse, Coline Delmas, Claudia Folco and Karine Goglio-Primard

What is the driving force behind the creation, animation and steering of communities of practice in organizations today? On the tools side, enterprise social networks are present in more than 80% of the top-40 companies and in 36% of companies in France. On the human side, individuals seek to find and interact with peers and experts within their own organizations to solve business problems or share common interests. Business leaders are investing in enterprise social networks[1] because they recognize the potential gains in employee productivity and want to leverage communities to support transformations in the organization, and improve efficiency through the sharing of best practices within and between countries or entities.

Industry leaders are noticing that the ways people make "educated" decisions in the workplace is evolving. Content gathering processes are characterized by some of the following phenomena:

[1]Enterprise social networks are tools such as Microsoft Yammer, Facebook Workplace or Slack.

- A distributed, mobile workforce is bombarded with information but needs a fast and efficient way to navigate through the data (identify relevant people, content and expertise) in order to perform their daily tasks efficiently.
- Global companies whose resources are scattered across geographies and time zones struggle to generate just-in-time and effective communication.
- New collaborative technologies in the cloud are providing communities with effective connectivity and collaboration solutions.
- Traditional network drives, storage systems and flat intranets are being viewed as inefficient and, as a result, are underutilized.
- As volumes of email and meetings increase, a simultaneous decrease in productivity is occurring.
- Organizational emphasis on only management personnel as a source for setting policy is now shifting to more effective involvement of the rank and file workforce.

Schneider Electric has taken these technological and societal changes into account and has developed strategies since 2011 for implementing communities of practice by leveraging its enterprise social network.

In this chapter, we identify good practices in community management at Schneider Electric through the implementation of the Communities@Work program. We analyze the key factors for the success.

We demonstrate that involving and putting members and facilitators of communities of practice in charge of its steering is an essential condition of success.

I. The History of Communities@Work, an Enterprise Community Management Program

1. *Collaboration example*

What defines the success or failure of a community of practice? The global community of client solution project managers provides a good example. In this case, a project manager based in a Gulf state was attempting to respond to a customer requirement. The project manager decided to analyze a utility customer's proposal for the creation of a mobile electricity substation. As most of the public knows, substations, for the most part, are fixed to the ground and comprise heavy, immobile transformers, wires and switchgear. This particular project manager decided to look to a community of internal Schneider Electric employees for help on how to design, assemble and implement such a mobile solution.

The challenge for the project manager was how to create a complete technical solution that would work. The first step was to identify a viable partner with experience in the area of trailer/container construction. The project manager decided to reach out to a wider audience within his community by posting his requirement in the forum of his community on the Schneider enterprise social networking tool called SPICE. One member of the community took the initiative to extend the reach of this request by cross-posting the message to other communities, including a purchasers community.

Within 24 hours of posting, three responses came in from worldwide community members. Within a month, a total of 25 responses had streamed in. Experts began to share their past experiences with various container construction vendors. More than 13 references of manufacturing facilities capable of undertaking such a construction project were posted on SPICE. Innovative suggestions on sourcing of specialized equipment and materials were proposed.

EXAMPLE OF COMMUNITY EXCHANGES ON SPICE

For the project manager in question, his knowledgebase on mobile substation solutions was immediately expanded. He gained education on both in-house and partner design approaches. He quickly developed a database of suppliers for needed materials and also established a checklist of best practices for mobile container design. Additional educational resources were shared as a series of posts grew into more substantial one-on-many conversations.

Based upon the collaborative advice received, the project manager and his team got a deeper understanding of the marketplace. Links among the various experts were reinforced. A solution was built based on the utility customer requirements and shared with the community. Without the help of the members of the communities, this new solution could not have been built so quickly with several innovative conception designs. The customer's issue was addressed and now Schneider Electric has a new, innovative solution that can be offered to new customers.

This emblematic example of collaboration, one year after the start of the Communities of Practice program and six months after the launch of SPICE, was a revealing example of the value of the community and the importance of an enterprise social network. While the number of communities and users of SPICE were still small, this testimonial was revealed to Schneider Electric executives: communities of practice bring tangible value to the company, its customers and employees and they are the best place to learn from others (Wenger-Trayner and Wenger-Trayner, 2020).

Therefore, these executives set as an internal objective by the end of 2014 to create 30 new active communities to help lead collaboration in critical areas of Schneider Electric's business. The reason for this objective was to break down silos, foster collaboration within the day-to-day work and increase visibility and participation.

Fulfilling this new mandate required a change in corporate culture. The following four principles underlie this change:

1. Collaboration begins with sharing.
2. Sharing begins by giving time, knowledge and expertise to others in the company.
3. Information that is not shared is lost.
4. Don't reinvent the wheel! Use what has been done… and build upon this base.

Schneider Electric has therefore decided to focus its efforts on a community of practice approach, internally called Communities@Work, to lead the thinking and the activities that encompass a concept called "communities for our collective intelligence". In 2015, it was decided to position communities of practice as "the best place to learn". Schneider Electric executives decided to move from the concept of unmanaged communities to the concept of community steering. Steering will only be successful if it is carried on by the members themselves. Indeed, involving and putting members at the center of this steering is an essential condition for its success. If members were not at the center of the steering process, they would see it as a control of their interactions and sharing of knowledge within their community. This hierarchical control would completely block the spontaneous sharing of knowledge among community members. The real challenge for the steered communities is therefore to reconcile management's desire to measure its results while leaving the communities free to act. Indeed, on the one hand, the company needs to measure the value provided by communities. On the other hand, the community only works well when members act spontaneously and are free to do what is right for them, without any intervention from the hierarchy. We will see that reconciling the wishes of the hierarchy and the will of community members can only be done by the members themselves who express their perception of value.

2. *Birth of the Communities@Work Program*

An enterprise community program is difficult to implement, because it is necessary to balance a top-down approach from management, a bottom-up approach from the field, an approach by functional/operational entity, a demand for tangible and rapid results, and the ignorance of the new transversal concept of community of practice — a pilot is required to test the concept.

In January 2011, Louis-Pierre Guillaume, a project manager with solid experience in Knowledge Management, just arrived in the company, took the initiative to look for existing communities of practice, because he knew from experience that communities improve collaboration. He identified about 20 communities of practice, survivors of a previous enterprise community project that had lasted from 2003 to 2007. The leaders of these communities were enthusiastic about the idea of being federated again under a "community management" banner.

In July 2011, Louis-Pierre presented a plan to three members of the Executive Committee (IT, Human Resources & Strategy), which advocated communities as a means to increase collaboration within the organization. The pilot project run between September and November 2011 was successful. In January 2012, they decided to launch an enterprise community program called *Communities@Work*. Moreover, an internal global survey in the first quarter of 2012 revealed that only 35% of employees of Schneider Electric considered that "collaboration is going well between the teams and entities". Results were needed, quickly.

3. *The Community management program*

Once the pilot was validated, the demand from the field established and the top leaders convinced, it was necessary to formalize the community program in order to encourage middle management to support the communities, or at least not to hinder them. A strong signal from the CEO, communication from headquarters, and appointed champions in the entities are useful to formalize such a program.

In 2012, Schneider Electric President and Chief Executive Office Jean Pascal Tricoire posed the following rhetorical question to all employees through the community program: "What if Schneider knew what Schneider knows?" In posing this question, he hoped to generate thought and awareness around the following important issues:

1. The drive to increase internal sharing and collaboration.
2. The need to break down organizational silos.
3. The initiative to increase use of new technologies to share information.

Louis-Pierre Guillaume was helped by a project team composed of managers whose entities allocated 5% of their time to deploy the program inside them. The main challenge was to identify potential communities of practice and convince people that these communities of practice could bring tangible value to employees, the company and its customers. They deployed a community of practice framework through which new communities could be created, whether at the initiative of management or employees.

The programme's objective for the end of 2014 was to create 30 new active communities of practice to help drive collaboration in the critical areas for Schneider Electric (R&D, Sales and Solutions).

In April 2012, one of the first challenges for the project team was to distinguish between a project team, a community of interest, an organizational team and a community of practice. Unlike organizational teams imposed by the hierarchy, a member's participation in a community is voluntary and without coercion from the hierarchy. Unlike project teams, the community has no time limit. In June, two communities were ready to be launched, their leaders and sponsors trained, their charters accepted. Other communities were being created and the team estimated that about 20 would be operational by the end of the year. As community members were scattered all over the world, it became clear that an enterprise social network was needed so that they could easily share regardless of their location or time zone.

Type of community	Definition
Community of practice (CoP)	Communities of practice (Wenger, 1998; Wenger *et al.*, 2015) are groups of volunteers who share a common practice. They seek to develop and accumulate knowledge about this practice. Throughout their history, their members create a shared directory of resources. They have a common passion without time constraints, without constraints imposed by their hierarchy.
Community of interest (CoI)	It allows members to be informed on a theme or topic (intelligence) and exchange ideas. Everyone uses the information for their work. This type of community evolves according to the needs and perceived usefulness of the information. It is often associated with a discussion group on an enterprise social network.
Project team	Project teams bring together members from different hierarchical organizations. They are under the explicit supervision of the hierarchy and have common objectives and time and cost constraints.
Organization team	Organizational teams rally around an operating division, function or entity of the company. These groups are often imposed by the hierarchy and their operation is continuous, until the next reorganization.

At the same time, Schneider Electric's IT department had chosen a tool for the enterprise social network (ESN) and was planning its deployment. Members of a community of practice that are spread over several sites can use a discussion group (also called a forum) in the ESN to facilitate their interactions.

At the end of 2012, 20 communities were launched after one year, a remarkable number since the objective was to create 30 communities in 3 years. We will explain how Schneider Electric communities of practice work and give examples from three different communities.

4. *The way Schneider Electric communities of practice work*

The communities of practice at Schneider Electric follow the Communities@Work framework and its key success factors: members who have a culture of sharing and connect their knowledge to each other, common objectives, and interactions through common tools that they adopt or create throughout their history. Members like to talk to each other because they are passionate about their domains. The passionate interactions they have with other members are favourable to creating a strong trust relationship. They exchange selflessly, simply for the pleasure of sharing what they know with people with whom they have created a trusted relationship. Collaboration cannot be imposed; it becomes a habit when people know and trust each other. The communities of practices are places where people can meet, exchange and build this mutual trust. Members set common objectives together that make sense for them.

For members of Schneider Electric communities of practice, the first role is the community leader[2] who leads the community's activity, stimulates and maintains its dynamism and vitality. They also encourage collaborative efforts. Between 10% and 20% of their working time is devoted to the community. Let's take as an example the typical week of a community leader. Every day, she will go to the SPICE discussion group of her community to relaunch ongoing discussions, answer questions, share information ... She will also organize the next webinar by writing the agenda, contacting potential presenters, sending invitations ... If she identifies information or documents that may be of interest to the entire community in the long term, she will then access the community directory (in the form of a folder in the cloud[3] or on the community's web page on the intranet) and store the information there to make sure it will be available to everyone at any time.

The second role is the core team, the local correspondents of the community leader in the different sites where the members are present. The core team supports the community leader in leading and making decisions for the community. Each member of the core team devotes approximately 2–5% of their working time to the community. Depending on the community, the role of the core team may vary. For example, the core team can be made up of contact people for each of the topics discussed in the community. It can also be an operational support for the leader. For example, one member of the core team will analyze the activity, another will moderate the forum in the ESN and another will organize the webinars.

The third role is the community sponsor, who supports and promotes the community. This person, usually a vice-president or department director, ensures that resources are allocated according to the needs of the community and encourages knowledge sharing by creating a supportive

---

[2]The term community manager is not used, because of the potential confusion with a manager role. A community leader may be a manager, or not (in Schneider Electric, 70% are not). A synonym is community facilitator.

[3]Cloud file storage are services provided by Box, DropBox, OneDrive, GoogleDrive, etc.

environment. The sponsor is a "bridge" between the community and the rest of the formal organization. He spends about 2% to 5% of his working time on the community. The concrete actions of the sponsor will be, for example, to validate the community charter and sign it with the community leader. The charter is the reference document of the community, because it defines the community, its objectives and the way it works. Together the sponsor and the leader will comment on the activity analysis, will decide on the main themes to be addressed and on the priorities. The sponsor has also as a mission to give the key messages to the members. For example, it is recommended that he participates in Kick-off meeting to explain the strategic objective related to the creation of this community.

Throughout their interactions, community members create knowledge and good practices that are stored, after validation by the leader or the core-team, in the community's shared directory, accessible from the community's intranet and the community newsletter.

The community also has an animation plan that is communicated to members. A schedule of activity or updates is proposed. A community can, for example, organize daily discussions on the ESN, monthly webinars, a monthly newsletter, an annual seminar, and update its intranet page every week.

EXAMPLE OF AN ANIMATION PLAN

Three typical Schneider Electric communities of practice are presented as follows. The Schneider Electric community framework is fully in line with the Wenger community of practice model. Communities of practice in the latter model (Wenger, 1998) are characterized by a mutual commitment of their members (ability of members to connect their knowledge with the knowledge of others), a common purpose (common objectives) and a shared repository of resources that bring together the tools that facilitate interactions between community actors.

5. *A technical community: Thermal & fluid dynamics M&S community*

It was created in 2008 and has 70 members. Of these, 16 are recognized Edison[4] experts. The Edison expertise program is highly developed in the R&D and supply chain entities. A jury awards three Edison grade levels

[4]Edison is a Schneider Electric expert recognition programme and technical career path.

to experts who meet specific criteria: innovation, invention and connection with the outside world.

This community is composed of experts with an R&D culture. They have a strong ability to connect their knowledge with the knowledge of others (mutual commitment of members). The common objectives of the community and its members are to help and coach the fluid dynamics specialists, through methodology and good practices in R&D projects and product and solution development. Through this community, best practices are deployed across operating entities and design centres, for example, on how to use both digital simulation tools and feedback on products installed at client facilities.

Community members have developed a shared directory of resources through their intranet site, containing each member's profile and skills, as well as shared documents, all of them accessible to non-members.

A community about sales process and practices: Sales Excellence Community

This second community of about 250 members is made up of managers in the domain of sales or marketing. It brings together a network of people who are in charge of sales processes in national and international operational entities, and who have a direct impact on sales practices and methods (Sales Excellence Directors, Sales Process Owners, Business Development Managers, etc.). These people have a strong ability to connect their knowledge with that of others (mutual commitment of members). The common objective of this community is to share and reuse best practices on continuous improvement of sales methods and approaches such as key account management, coaching, customer relationship, KPI calculation. This knowledge is also available to non-members. For example, the Italians have defined the frequency of interactions between salespeople and their managers with a list of points to be addressed. Members in the Gulf countries are cross-referencing vendor interactions with the analysis of the potential of customers, in order to know the most important customers. The English are organizing renowned sales training courses. The Spaniards are crossing customer satisfaction information with sales analysis, in order to know if they are satisfying the most important customers.

6. *A community about a market segment: Healthcare solutions community*

It brings together a group of people who share a common interest in the Healthcare solutions business throughout the organization. Community members are aware that they work and learn together to develop the business of this segment. Its members have a strong ability to connect their knowledge with that of others (mutual commitment of members). The common objective of this community is to capture, share, create and reuse health knowledge within sectors, regions, businesses, and global strategic accounts, knowledge that is also accessible to non-members. In 2013, the company won a contract in an Australian hospital. One of the criteria for awarding the contract was that the applicant had to demonstrate that it had access to a global network of knowledge and good practices. The hospital's executive director said: "Schneider Electric's international network allows us to receive the best technological developments from all over the world". The community sponsor, Vice President of the Healthcare Solutions market segment, said: "We could not bring value to the customer without the community. Tacit knowledge is the knowledge that has the most value; it enables us to offer solutions that make a difference, shared by key account vendors and solution architects."

These three examples allow us to better understand the benefits of communities of practice for members.

Communities of practice	Benefits for the members
Thermal & Fluid Dynamics M&S Community	Methodology and best practices (R&D and product and solution development projects). Good practices on how to use digital simulation tools and on feedback on installed products. A shared directory of resources (shared technical documents). Access to community members and their expertise (Edison experts).
Sales Excellence Community	Good practices (sales methods and approaches: key account management, coaching, customer relations, KPI calculation). Access to community members in different countries to access their experiences and expertise: • Measure the frequency of interactions between salespeople and their managers. • Cross-reference salesperson interactions with customer potential analysis. • Reputable sales training. • Cross-reference customer satisfaction information with sales analytics.
Healthcare Solutions Community	Access to a global network of knowledge and best practices. Access tacit knowledge to offer differentiating solutions to customers.

7. *Key success factors of an enterprise community program*

Three main key factors for the success of the Communities@Work program can be highlighted:

1. Follow the same principles as the communities it supports.
2. Measure the value of the community as expressed by members.
3. Serve the communities and implement a management system inspired by the needs of the communities of practice.

A program guided by the same principles as the communities it supports

Communities aim at gathering knowledge around their common practice. Strong principles are supported by communities according to the three examples presented in Section 2 of this chapter:

- Communities encourage knowledge and best practices sharing.
- Members connect their knowledge and expertise with others, inside and outside of the community they belong to (knowledge open to non-members).
- Knowledge is capitalized and can be re-used in the daily activities of the community (best practices, R&D projects, solutions selling, marketing studies …).

The Communities@Work program promotes these principles.

8. *Measurement of the value perceived by communities' members*

The added value of communities has to be expressed and measured by their members. It cannot be measured from an external system built by the management, as most of the KPIs of the company's activity are — as per the example presented at the beginning of this chapter — where the members of the worldwide community of project managers have collaborated on a project for a mobile electrical substation. In this case, community members themselves express the added value of the community: they created in a short time an innovative solution from a complex problem which can be proposed to other customers.

Why should the added value be expressed by members? Because we are facing a dilemma. On the one hand, the community needs to measure communities' added value. On the other hand, communities work better when members act spontaneously and when they are free to do what they want, without management guidance. A measurement system, externally imposed, would be perceived as an intrusion in the community life: members would feel deprived of the community dynamic and would lack motivation. The solution is to measure the value expressed by the members themselves.

Formal metrics produced by each community contribute to demonstrate this added value. For some community leaders, this value can be represented by the number of posts shared per week on the community's group in the Enterprise Social Network (ESN). For others, it can be the participation to community's webinars (participation and exchanges). Schneider Electric decided to ask the members directly through an annual global survey, without going through the leaders' filter, but with their validation.

Since 2013, Schneider Electric measures members' satisfaction and promotes communities through an annual worldwide campaign, the *Active Community Label* (Figure 1). This campaign aims at giving a prize (the Active Community Label), in order to recognize the most active communities, the ones providing the most added value to the company, to its members and to customers, according to the members' voice. The campaign provides other qualitative information per community, such as the type of added value (the community brings more business, helps to save time, reduces cost, etc.) and members' suggestions for improving the community. Only communities that follow the model (a leader, a sponsor and an updated charter in the year) and that are at least 6 months old can participate. Community leaders register for the campaign if they feel that their members are active enough to share feedback.

This label presents multiple benefits. It provides recognition to winning communities, to their leaders and sponsors. It increases members' engagement because members are being recognized in their community, and outside by their peers. It inspires emulation among communities to win the label. Non-winning communities are being encouraged to review their strategy in order to win the label the next year. The label provides visibility to the Communities@work program and it encourages other communities to join the program. It boosts communities of practice and their members as it is based on members' voice, through a questionnaire.

Figure 1: Active community label 2018.
Source: Schneider Electric.

Communities' members assess their community for themselves, their customers and their company.

In the latest campaign in 2018, 68 out of 220 communities participated. The other communities could not participate either because the leader was not ready or did not want to, or because they did not meet the criteria. 2,900 voters (a representative sample out of the 10 742 respondents) elected 29 communities of practice as active, based on the responses of the question "I consider that my community is an active community, because it provides tangible value to me, my business or my clients" (Figure 2).

A global score is calculated from that question, the Net Activity Score (NAS). This score is the weighted average of the four values. Its value ranges from +100 if everybody marks "strongly agree" to −100 if everybody chooses "strongly disagree". This score increased continuously until 2016, indicating a global increase of the value perceived by members. Then it dropped in 2017, causing a confusion among the few community leaders who saw a big drop in the measurement of the value of their community. This drop could be explained by a possible complacent attitude of some of them, who had been leading their community for a while. Members, feeling that the energy and passion in the community was not as it used to be, reacted and gave the thumbs down. Following that *annus horribilis*, the leaders apparently reacted positively, because the NAS went up the following year.

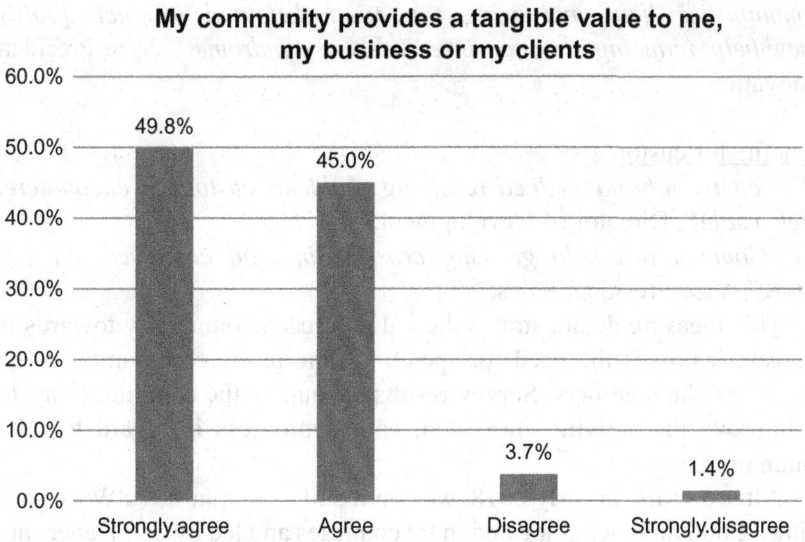

My community provides a tangible value to me, my business or my clients

60.0%
50.0%
40.0%
30.0%
20.0%
10.0%
0.0%

49.8% Strongly.agree
45.0% Agree
3.7% Disagree
1.4% Strongly.disagree

Figure 2: Aggregated results for all the communities in 2018.

Source: Schneider Electric.

	2013	2014	2015	2016	2017	2018
NAS	56	61	68	74	67	69

After the campaign in 2015, sponsors of winning communities have been surveyed. The following statements summarized the three types of benefits.

<u>Benefits for members</u>

"Most of the top performers in our activity are the most active members of the community", Activity Director.

"The community is not a gadget any more. It provides guaranteed responses to its members in less than 48 hours. The members have vastly increased their competencies", Vice-president Human Resources.

<u>Benefits for the company</u>

The communication and sharing certainly assisted the entire community to drive improvements in Quality and Service, Efficiency and Productivity, Inventory, and Safety", Vice-President Logistic.

Communities help increasing time-to-market and product quality. They help reducing the non-invented-here syndrome", Vice-President Innovation.

Benefits for customers

"The community has helped resolving problems customers encountered after a sale", Director of Development.

"Communities help growing cross-selling, all countries, all segments", Vice-President Sales.

This measure demonstrates the value of each community towards its sponsor, it proves the need for spending time in the community for the leader and the members. Survey results encourage the community leader to improve the activities provided, and to progress in regard to other communities.

After 6 years, in early 2018, we count 220 Communities@Work gathering 32 000 members,[5] located in 90 countries and led by 250 leaders and co-leaders.

9. *A programme that serves communities of practice*

The role of a community of practice program like the Communities@Work is to support community leaders and their members in achieving their goals. It is not a program to control communities but rather a real support program for these communities.

Thus, this program is dedicated to help any Schneider Electric employee that wishes to create a community or that is already running one.

We present below the good practices the Communities@Work programme has implemented to bring its support to communities:

Get hierarchy support for the Programme and Communities

- Establish this programme as a global company programme in order to make it visible, thanks to the broadcast of the internal communication network. Identify a full time *Knowledge Management Officer (KMO)* to run the programme.

[5] 83% of them are members of one community only, 13% of two and 4% of three or more.

- Convince an executive committee member to sponsor the programme in order to bring legitimacy with the hierarchy and the communities of practice (leaders, members …).
- Convince managers to give their team members dedicated time to participate in the life of their community of practice (5%), even if and especially if their time is billable.

10. *Help and advise new communities of practice creators*

- Define a community of practice framework, like the Communities@ Work, with basic principles (leader, sponsor, charter …). Principles that guide those who wish to create their community of practice.
- Spend an hour (the KMO) face to face or virtually with the new community leader to go through the fundamentals of a community (the notions of group, shared objectives and interactions). Also, ensure regular follow-ups to support the leaders in their mission. We are talking here about coaching on-demand. The leader can have some coaching from the communities of practice programme managers.
- Provide tools to take advantage of the global Enterprise Social Network (ESN), deployed to all employees, to foster collaboration and cross-functional exchanges. If the ESN does not exist, launch it at the same time as the communities of practice programme.
- Offer training on collaborative tools throughout the year, for instance, "how to organize a webinar", "how to leverage Klaxoon (a collaborative tool allowing interactive meetings)"…
- Help communities (*Active Community Label*) to measure their business value and publish success stories in order to justify the time spent in the community, both by the members and the leader. This measure is important for members and leaders to be able to evaluate the dynamism of their community and to be considered legitimate inside and outside of their community by their peers.

11. *Give a status to community leaders*

- Ask leaders to dedicate time (10–20%) to run their community, which has a direct impact in encouraging interactions and allowing the organization of dedicated events and thus facilitating community members' engagement.

- Give a place to the community leader role and bring them visibility. For instance, each year, the programme team encourages managers of community leaders to include their role as community leaders into their annual goals. Events are also organized to widely recognize their job and the value their community brings.
- Create a Community Leaders' Network to offer the possibility to share with their peers about community management topics. This last point will be detailed in the following section of this chapter.

II. The Community Leaders' Network

In this section, we will talk about the Community Leaders' Network. Each person taking over the role of community leader becomes *de facto* a member of the community leaders' network.

1. *History of the community leaders' network*

Mid-2014, with a two-years perspective, Louis-Pierre Guillaume realized that although community leaders were becoming more and more empowered in their role, new leaders were beginning to replace the old ones; which created a new need to learn from peers to be more comfortable in their role. Thus, the Community Leaders' Network was created in 2014, in which community leaders are themselves members. This network is run by the same two programme facilitators (a full time Enterprise Community Manager and a 20% KMO), sponsored by the two same sponsors as the programme (Chief Digital Officer, CDO and Senior Vice-President of Learning), with an active team of 17 volunteers (core team) made up of community leaders and sponsors in 5 countries. The network has its own animation plan, as every community.

The creation of this network allowed the Communities@Work programme managers to have a better dialogue with members of communities and to better escalate the needs of the field. It is important to note that a top-down approach is necessary at the beginning to launch the community programme and promote its value. A caring co-management is the next step. The ultimate goal will be to have communities as an integral part of the corporate landscape, known and recognized at the highest level, so that community leaders can fully manage the network themselves without the help of a central team. However, reality shows that today the central team still plays a key role in enrolling new leaders and being agile in a fast-paced organization.

2. *Scope, purpose and benefits*

The Community Leaders' Network promotes mutual respect and support among its members. It is an environment where mistakes are learning opportunities. There is no taboo and no bad questions either; the community is engaged to answer questions in less than 24 hours. For instance, a new section has been created in the webinars: "I have a question". During this section, a community leader takes the floor and explains a problem or a challenge he has encountered with the community; during the webinar other community leaders propose some solutions. The discussion then continues in the Yammer group of the Community Leaders' Network (Yammer, our ESN, Enterprise Social Network). If someone asks a question, they want an answer. The later the answer comes, the more people will lose interest in the community because they finds it doesn't bring any concrete outcome, and therefore they will connect less and less to the ESN. It is a vicious circle that ends up killing a community. Guaranteeing a response in 24 hours to keep this attraction is a challenge, as this implies that the Enterprise Community Manager, and members to a lesser extent, are on alert, on the ESN. It also implies that, if there is no response after 12 hours, the Enterprise Community Manager must contact directly people who may know the answer to ask them to write it down on the ESN. It is essential to push the use of the ESN rather than email, because the answer will be visible to all, so it may be useful to other people than the requester.

The network is made up of 250 members, located in about 20 countries, and leading 220 communities; some communities have two leaders, a senior and a junior, to reduce workload of each and use the strengths of each (e.g. knowledge of the field vs ease in the use of collaborative tools).

The Community Leaders' Network organizes an annual event to reward communities that bring the most tangible value, escalate good practices back to management and push employees to join communities.

Once the structure of the Community Leader's Network animation plan has been defined, the Enterprise Community Manager and the core team of the network need to imagine the content. The sources of inspiration are multiple.

The sponsor of the Community Leader's Network gives guidelines and the priority of the year, members express their needs, external stakeholders are regularly invited to bring new ideas and fresh air. But the most inspiring source comes from the community leaders themselves. Every other month, during the webinar, a community leader describes his community and its activities. For instance, in March 2016 the Change Leadership Community Leader showed the commitment of each member of its core team to run the community, the way to make a global follow-up of the activity of the members of the community, and the typical format of its webinars. He was chosen to present because he was a volunteer and his community is known to be active.

The Enterprise Community Manager (facilitator of this network) observes activities in each community and promotes the best ones. Members ask for facilitation tips and tricks. They learn from each other and realize that the role of community leader is not to be a knowledge provider, but to be a bridge between knowledge offer and demand. For example, a working group was set up in January 2018 of four volunteers, to think about how to involve the sponsor, the core team and the top management in communities. After two months, the recommendations were shared with other community members during the webinar. The Enterprise Community Manager role was just to suggest a framework, theme ideas, dates and publish the result.

Examples of community leaders' profiles

Community leaders behave differently in the animation of their community and in their participation to the Leaders' Network. The different configurations of community leaders' profiles demonstrate the diversity and the creativity of each community. According to our observations, profiles can be grouped into six categories.

Profile 1: An experienced community leader, strongly involved in the Leaders' Network

This person is strongly engaged in the Leaders' Network. He belongs to the core team of the Network and takes part in the decision-making process (yearly planning, main events …). As part of the core team, he is a role model: he shares testimonies in webinars, he participates in all the activities and shares his best practices linked to community facilitation. He captures, shares and creates information and knowledge related to the community's main domain. He gathers experienced members to solve common issues in this domain. His animation plan is precise: frequent webinars, opportunities for top management to join and share their insights with the members, ad hoc training, on-demand coaching, tips and tricks, best practices from universities and other companies. He leads the community for at least 5 years. His community possibly wins several times the Active Community Label and the Learning Community Label.[6] His activities as a community leader and his day-to-day job are not related.

Profile 2: An experienced community leader, with a limited participation in the Leaders' Network

Her participation to the social network is low and she is not engaged in the activities of the Network. However, her community is efficient, and it wins the Active Community Label and the Learning Community Label. The success of this community depends on the role of the community leader that is fully integrated with her day-to-day job, with a strong involvement of the sponsor. This example is interesting as it shows that some people, with transversal organization and communication skills, supported by the management, with a legitimate position among peers, can perform on their own. This community leader spends 10–20% of her time to manage the community. If she would like to participate more in the Network, she would have to spend more time in the community, which she cannot.

Profile 3: A non-experienced community leader, strongly involved in the Leaders' Network

This active member in the Network is a newcomer in his role of community leader. He wants to learn the tips and tricks, best practices, to help him manage his community. He needs support to properly launch his community. By joining the Network, he gets support from the Community programme Manager but also from other experienced community leaders. He participates in most of the activities of the Network and sometimes helps with some activities.

Profile 4: The silent member in the Leaders' Network

She is unknown in the Network because she does not show any signs of activity. Her community is more or less active. She does not have time to

[6]The Learning Community Label is given to communities that reach a high score for the question "I am learning from the members (through webinar, direct conversation, ESN exchange …) of my community."

spend in the Network and does not see the benefits. This type of community leaders is hard to catch and it would require spending some individual time to convince her of the added value.

Profile 5: The full-time community leader
Community management is his full-time job because his manager is convinced that his role will develop more business. His community gathers many employees in a cascading structure, as it supports the adoption of one tool in the company. This leader is strongly involved in the Network and shares a lot of ideas.

Profile 6: The senior-junior duo
The senior community leader has more than 20 years of experience in the company. He is recognized by his peers and enjoys communicating. Nevertheless, he cannot spend 20% of his time to manage the community. Therefore, he decided with his sponsor to hire an intern to support him in the webinars' organization, running the animation plan. The junior community leader could replace the senior leader when he would move to another position.

Leaders' profiles depend on the community leaders' conditions: innate talent in animation, management support, priority of this activity — time allocated to do it, link between leader's role and his job. A leader can move from one profile to another due to changes in the conditions. This proves that there is no correlation between the leader activity in the Network and the level of activity of his community, as measured by the Active Community Label.

The 220 communities of the programme today are run by 250 leaders and co-leaders, some of them have this senior–junior duo profiles.

III. Lessons Learnt from this Diversity of Community Leaders' Profiles in Schneider Electric

In 2018, from HR system data, we analyzed the leaders' profiles. 250 leaders, mainly men (2 out of 3), animate the communities of practice in Schneider Electric. Leaders have a professional and mature experience (67% of them are part of the X generation) which legitimizes their role. They master the subject of the community and they put in contact members. In many cases, millennials support experimented leaders, in a co-leading role. They tend to be more comfortable with Enterprise Social Network and digital tools. They represent a strong ally in the animation of the community.

Community leaders are mainly located in France, United-States and Spain (respectively 43%, 14% & 13% vs 14%, 14% & 4% of connected employees). They tend to be closer to central teams or hubs.

In 2018, out of the 250 leaders, only 23% of them were managers — compared to 45% in 2015. This change can be explained by an internal transformation, reducing the levels of management — and thus the number of managers — to obtain a flatter organization. We see also this change as an emergence of a new form of management as facilitator, being a community leader and animating a network of people, to break silos.

IV. Conclusion

Our analysis highlighted three key success factors for a programme of communities of practice. First, the programme must be guided by the same principles as the communities it supports. Second, the measure of the communities' added value must be assessed by its members. And last, the programme must serve the communities. Its directions must be inspired by the needs of the communities of practice. Through the third key success factor, we identified the governance's best practices: get support of the management for the programme and the communities (a strong sponsorship among the executive committee), provide help and guidance for creation of communities (community framework Communities@Work, a dedicated central team, coaching and training, proof of the tangible added value with the Active Community Label), and provide a status for community leaders (Community Leaders Network).

To successfully launch a programme of communities of practices in a company, the following steps are recommended:

1. **In the coming weeks:** Start to plan a roadmap. Identify benefits that the organization would be perceiving following the launch of the communities.
2. **In the next six months:** Identify a first community that would require low investment but that could produce positive results in the short term (a community focused on a specific domain, with a motivated

leader, a members' list and a business sponsor). It would be used as a pilot. Members would enrich it with their contribution.

3. **In the next year:** Identify domains where communities of practice could provide more value. Identify a high-level sponsor for the company programme and define together the scope, the budget and the resources of the programme.
4. **In two years:** Build a long-term plan for the programme and launch other communities. Create succession plans for community sponsors and community leaders. Deploy a measurement programme for long-term tracking.

In the context of complex organizations in a fast-paced environment, we would also like to present some food for thought, some areas that we believe are worth taking a deeper look at as they will definitely change the landscape of communities of practice in the future:

- How to strengthen the interactions between experts (technological or other fields such as IT, industrial …) and communities of practice.
- Explore the work of communities of practice in terms of innovation proposed or produced by them.
- Questioning on the hybridization of the traditional business model (silos) and communities of practice.

References

Wenger, E. (1998). Communities of practice. Learning, Meaning, and identify, Cambridge University Press.

Wenger-Trayner, E. and Wenger-Trayner, B. (2020). Learning to Make a Difference: Value Creation in Social Learning Spaces. Cambridge University Press.

Wenger-Trayner, E., Fenton-O'Creevy, M., Hutchinson, S., and Kubiak, C. (2015). Learning in Landscapes of Practices: Boundaries, identify, and knowledgeability in practice-based learning, London: Routledge.

White paper of Schneider Electric: Wanted: An Active, Viable, Collaborative on-Line Community. http://www.guillaume.nu/documents/Online-Community-Schneider-Electric-2014.pdf.

© 2021 World Scientific Publishing Company
https://doi.org/10.1142/9789811234286_0005

Chapter 5

Key Success Factors for Communities of Practice in Innovation: The Case of the Groupe SEB

Lusine Arzumanyan, Charlotte Wieder and Claude Guittard

This chapter is based on the study by the Groupe SEB, a French multinational company specialized in small domestic appliances, of its Community of Practice (CoP) in innovation. Innovation is at the heart of the Groupe SEB's strategy. It is a key element in its differentiation from competitors, and a lever for growth. In order to maintain a high rate of innovation, the group invests nearly 3.5% of its turnover in R&D activities, and each year launches more than 300 new products. Various key products have influenced its development: these include the *supercocotte* in 1956, the oil-free fryer (Actifry) and the FreeMove (wireless iron). The company's secret weapon: its CoP for Innovation, which brings together nearly 1,300 members from Marketing, Design and R&D involved in the Group's innovation processes.

The creation of a CoP for innovation provides an answer to the need to generate and accelerate the development of new ideas internally, in order to meet new consumer expectations. Through this approach, the Groupe SEB intends to take advantage of the synergy created by interactions between various stakeholders, with the aim of developing a proactive, dynamic context to foster innovation.

Based on the experience from/of this CoP, we provide the readers with the answers to the following questions:

— What are the challenges to be met in the process of setting up a CoP for innovation?
— What are the key success factors observed, whether in CoPs administered by a community manager, or non-administered and therefore self-managed by their members directly?
— What are the benefits of a CoP?

To this end, we firstly present the history behind the creation of the Groupe SEB's CoP for Innovation, and then discuss the challenges to be met during its implementation. Finally, we present the key success factors identified through our observations, before concluding on the various advantages provided by the CoP.

I. History of the Community of Practice for Innovation at Groupe SEB

1. *The Objectives*

In 2011, following its previous business plan, the Groupe SEB's General Innovation Department decided to implement a Community of Practice (CoP) for innovation, with four objectives (cf. Arzumanyan and Mayrhofer, 2016):

(1) The development of cross-functionality between its three business units. The challenge was to improve the sharing of projects in progress, or which had already been completed.
(2) Stimulating and accelerating innovation, using improved transparency in the dissemination of information.
(3) Sharing of monitoring activities, in order to improve the cross-fertilization of research efforts.
(4) Leveraging of existing knowledge, by means of a more advanced and cross-disciplinary cooperative assistance system.

The Groupe SEB's CoP for innovation was not created from scratch; as its innovation teams were already in place and there were already more or less structured informal networks, such as the CoP for electronics.

In addition, an annual event called the "Technology Forum" was organized every year, starting in 2001, by the Group's Methods and Tools for Innovation manager. The purpose of this forum was to facilitate exchanges between R&D employees. We can therefore consider that this event marks the real beginning of a coordinated initiative to share and pool knowledge.

The Groupe SEB's CoP for innovation is composed of several sub-communities, such as the "Sustainable innovation" sub-community, whose objective is to raise awareness and encourage innovation stake-holders to integrate more sustainability into their products and services, or the "Innovation tools" sub-community, whose objective is to facilitate the sharing of useful innovation methods and tools for innovation projects. The common feature of all these sub-communities is their vocation to contribute, each in their own way, to the acceleration of innovation within the Group. Whether this be achieved through better knowledge sharing, faster dissemination of information, or better mutual assistance. In the following section, we briefly describe the resources that were made avail-able to the Group's CoP.

2. *The resources deployed*

To design and implement a coordination system, the General Directorate of Innovation (DGI) recruited an innovation community manager in September 2011. With a professional Master's degree in information systems management and a doctorate in industrial engineering applied to innovation and agility, her mission was to organize and create a network among various actors, with the aim of promoting and facilitating innova-tion. It was thus necessary to design and implement a facilitation system based mainly on the organization of physical meetings and the implemen-tation of *What If*, the CoP's internal online social network for innovation.

Having presented the background of the Groupe SEB's creation of a CoP for innovation, we now present the challenges associated with the implementation of this CoP.

II. The Challenges Encountered when Building a CoP for Innovation

During the implementation of its CoP for innovation, the Groupe SEB faced several challenges and was able to overcome them due to its constant awareness of the needs of community members and to the adjustments it applied to the development strategy of this CoP.

The first of these was to understand the expectations of the community members, in order to propose a suitable coordination plan. One could also refer to the coordination of communities, with an "s", because this was instigated at two different levels:

1. In the innovation community (1,300 people), whose members have in common that they contribute to innovation projects during the upstream phases of the innovation process;
2. With all the sub-communities (approximately 30), which allow their members to exchange ideas on common subjects (expertise, monitoring themes, profession, etc.). Their size generally ranges from 5 to 150 members, in the largest sub-community.

1. *Organization of innovation forums*

From the very first days of her employment, the CoP manager was involved in organizing the Innovation Forum: the largest annual gathering of community members.

The Innovation Forum 2011 (an event similar to the former Technology Forum) brought 250 people from marketing, design and R&D together. One of the objectives of this forum was to promote transverse sharing and cross-fertilization between several business units. To achieve this objective, several activities and workshops, often led by experts from outside the group, were set up to create exchanges and bring together participants from different backgrounds. For example, in 2011 the members worked in small multidisciplinary groups on prospective usage scenarios related to the group's four strategic axes of innovation. The objective: to introduce them to a new working methodology, to imagine products and services in the fields of "Ageing well", "sustainable innovation", "health, beauty and well-being", as well as "Connected world & habitat". Since 2011, this event has become a biennial event.

At the 2013 Innovation Forum, the trades represented were the same as in 2011. However, this time the Innovation Forum had three objectives: (1) to contribute to the sharing of information on projects and good practices among community members, (2) to improve the capacity for innovation, breakthrough and concept generation and their implementation, and (3) to expand and strengthen the participants' internal network. It was thus very well matched with the challenges of improved transversality, transparency, pooling and capitalization within the CoP for innovation.

A knowledge fair was organized during this forum. On this occasion, members of the business units and innovation poles presented the Group's future innovations on stands (via posters and other resources), thus facilitating exchanges between participants and exhibitors. The advantage for the participants was to have the opportunity to see, contact and exchange with each stand manager. This forum focused mainly on the sharing of internal experiences.

2. *The implementation of innovation events*

Following consultation with those who could co-organize innovation events, an innovation event programme was elaborated and implemented. "This programme offered: *themes for highly diverse innovation events: involvement of a supplier to discover new materials, feedback on a tested innovation methodology (for example, innovation through usage, crowdsourcing, etc.), our strategic axes of innovation, or shared expertise such as prototyping and modelling*)". (Innovation Community Manager).

The aim of the innovation events was to supplement the Innovation Forum, with times for meetings and exchanges, often in smaller committees (see Box 1).

Box 1: Innovation event on sedentary nomadism or the nomadic use of products

An innovation event was organized on the theme of sedentary nomadism, or the nomadic use of Groupe SEB products in the consumer's home: *I take a product that is generally used in the kitchen, and use it in the garden or bathroom, with the usage-related constraints that this can represent* (Innovation Community Manager). The aim of this event was to allow battery experts (specialized in the storage and use of energy) to present their technological roadmap, and then in a brainstorming session with marketing, R&D and design experts, to imagine what usage scenarios could be considered, while taking into account the relevant technical limitations. *For example, I want to iron from any room in my house, without having to be next to an electrical outlet* (Innovation Community Manager). This event increased the participants' awareness of the fact that consumers are increasingly mobile within their homes, and that it is therefore necessary to offer products and services adapted to these practices.

Taking part in this type of event offers several advantages to the participants:

It is already a way of taking a step back from everyday work, leaving the office and focusing on an often completely different theme. This breath of fresh air generally allows an individual to "recharge his or her batteries" and feel more motivated when returning to work. Then, it allows the employee to develop his or her professional network, either by seeing acquaintances or meeting new people. This makes it possible to share work in progress during informal moments such as receptions or coffee breaks. It can also foster ideas for new shared projects. Thereafter, this can facilitate mutual assistance between members.

Even if these events are not necessarily devoted directly to a person's current innovation projects, they can indirectly contribute to these in the long run, through better mutual support, or the emergence of new project ideas.

This "new formula" for facilitation was also designed to increase the frequency of physical meetings, in order to establish an operational working rhythm in the CoP for innovation. It was also an opportunity to increase the likelihood of serendipitous events or "happy coincidences". *People need to meet each other [...],in order to make connections, it's easier when you bring them together in small groups focusing on common topics and interests* (Director of Innovation Processes).

These events have sometimes given rise to new CoPs, such as the prototype-model community following the event organized on this subject in 2012. The participants at this event were the proto mock-up experts as well as their "internal customers" (innovation project managers, marketers and designers). The aim was to present the resources of each workshop in terms of human and material resources, to present the know-how and specific expertise of each participant, to introduce examples of already completed initiatives, and to work with the buyer in charge of the supplier panel in an effort to identify areas for improvement.

Thus, many thematic workshops and CoP events were organized and implemented between the two Innovation Forums held in 2011 and 2013.

3. *The implementation of online social networking tool to facilitate exchanges*

The CoP's internal social network for innovation *What If* was launched at the end of 2012. Its name embodied the spirit required to innovate, asking the question: *What would happen if we explored new avenues for the Group?*

The aim of the CoP's social network was to facilitate the accomplishment of the innovation community's objectives, namely, more transversality, transparency, mutualization and capitalization. It thus had several goals:

(1) To allow exchanges to continue *via* a dedicated digital platform, between physical meetings such as the Innovation Forum;
(2) To facilitate transversal and informal exchanges between community members who are geographically dispersed over several sites in France (Ecully, Rumilly, Pont-Evêque, Selongey, Is/Tille, Vernon, Mayenne, Lourdes, etc.) or in international locations;
(3) To share the results of business intelligence among members, instead of simply sending them by e-mail to a few people, which would constrain the distribution of this information within the community;
(4) To promote informal information exchanges and coordination between members, rather than just storing validated information such as in the case of Product Lifecycle Management (PLM), SAP or Sharepoint.

Nowadays, the implementation of a virtual sharing tool is very useful for the promotion of community exchanges. Indeed, we now consider digital systems to be an essential component for the continuity of interactions between in-person events. This facilitates the exchange of coordination information, for example, to set a date for the next in-person meeting/face to face, as well as for the sharing of documents, images and videos. In this age of corporate social networks, there is an ever-increasing number of companies commercializing this type of tool. However, they are not essential for the implementation of a dynamic and high value-added CoP. Indeed, before the arrival of *What If,* some Groupe SEB CoPs functioned with a simple email distribution list. This was more than enough to: remain in contact, announce upcoming meetings, set the agenda, and share documents and other types of content. However, it is true that a professional social network at work, such as a private group on Linkedin, Yammer, Jive, Slack or Facebook, makes things much easier. In effect, it is more comfortable to share within an online space, which at a minimum will allow reference documents to be stored within a common library, and online conversations to be held *via* a microblogging solution. The walls of these social networks have the advantage of keeping a record of exchanges between members, and facilitate interactions between them with varying degrees of investment and commitment: from a simple *like* to show interest or indicate that a post has been read, to a comment that enriches the content with a member's own perspective, to a more developed post with an attached file, for example. It is up to each CoP to define the tool that works well in accordance with the applications for which it is intended.

4. *The emergence of sub-communities of practice*

Community support was provided at two levels: firstly by fostering sub-communities in *What If*, the CoP's collaborative tool, and then by setting up a network of community managers to lead it.

Before the arrival of *What If*, there was no steering of the CoP. The implementation of a collaborative tool provided an opportunity to implement and structure a guidance strategy for new or existing CoPs, in order to clarify the target objectives and applications, the associated benefits, and enhance the skills of community managers and support initiatives. Following the implementation of this collaborative tool, several sub-communities were created. Some of these corresponded to the Groupe SEB's strategic axes of innovation such as the *Ageing well, Connected world and habitat, Health, beauty, well-being,* and *Sustainable development* sub-communities. Others, on the other hand, emerged following the Innovation Events, such as the *Innovation by use*, *Crowdsourcing, Prototype/Modelling, Materials, Development Managers,* and *Calco-magnesium deposition* sub-communities.

The communities created in the *What If* context were:

— either sub-communities which existed informally but had not been officially identified, such as the proto-model community,
— or communities that had been created following the organization of Innovation Events, such as the *Innovation Tools* community.

The advantage of *What If* is that it works as a means of revealing pre-existing or potential communities. In addition, thanks to its directory-function, it allows each CoP member to observe existing communities and potentially follow them. This visibility of sub-communities also has the effect of encouraging the creation of new communities.

In the collaborative tool, each sub-community has its own shared space, with its own "wall" of conversations on which members can post and exchange, as well as its own library allowing each member to capitalize on the content created by other members, and to find documents with keywords.

To consult the content of a sub-community's wall or library, the user must "join" that sub-community. All existing sub-communities can be consulted *via* the Communities directory. If a sub-community is public, any person who is a member of the Groupe SEB's CoP for Innovation can join it with a click of the mouse. If it is private, the person must apply to join it. If it is secret, it is visible to its members only.

The dissemination of the creation of a new sub-community, whether public or private, occurs mainly by word of mouth. During a discussion around the coffee machine, a colleague tells another colleague, for example, that a question has been posted in a particular sub-community and that perhaps he/she should join it if he/she has the answer. This informal dissemination is also complemented by more formal email communications between all members of the Innovation CoP, whenever a new sub-community is launched.

To support the management of the various sub-communities, community manager training was offered to 25 people. They were then able to exchange ideas within a dedicated sub-community. The purpose of setting up this network of community managers was to boost exchanges within each sub-community, thanks to coordination relays assigned to each of them. Trained members were often already active contributors to *What If,* regularly sharing content or commenting on the publications of other contributors. The innovation community manager thus recommended that they go a step further in their involvement, by becoming the manager of their community, following appropriate training. A training module was created based on the relevant literature, the innovation community manager's experience and a consultant specialized in this field. Training took place over two half-days, in person, with approximately ten community managers on each occasion. The aim of the first half-day: to lay the foundations of the community, by working on the fundamentals (objectives, core group to be involved, target users, individual and collective advantages). The second half-day was devoted to daily coordination. In particular, with the definition of an activity plan to meet the objectives of the community, and role-playing games to help community managers discover the different situations they could face: establish a consensus with the core group, convince a person to play the role of sponsor, welcome a new member, etc.

To further develop their expertise, a one-hour exchange was held once a month: the community managers' meeting was arranged. The aim: share good coordination practices and further develop their expertise in coordination. Each month, a specific theme was proposed: presentation of the new functionalities of *What If,* administration of each individual's sub-community in *What If,* and feedback on a coordinating action that had been completed.

After briefly presenting the process of setting up Groupe SEB's CoP for innovation, we now discuss the key factors for the success of a CoP (managed or non-managed). These factors were identified through research work conducted with some 30 sub-communities from Groupe SEB's CoP for innovation.

III. Key Factors for the Success of a CoP for Innovation

After having studied the development of the Groupe SEB's CoP for innovation, as well as its sub-communities, we identified several key factors for its success (Figure 1).

In the following analysis, we first present the key factors for the success of a CoP for innovation, regardless of how it is created (whether managed or not), before focusing on the key factors for the success of managed communities. We refer to a *managed community* when a clearly identified initiative leader or community manager is in charge of leading

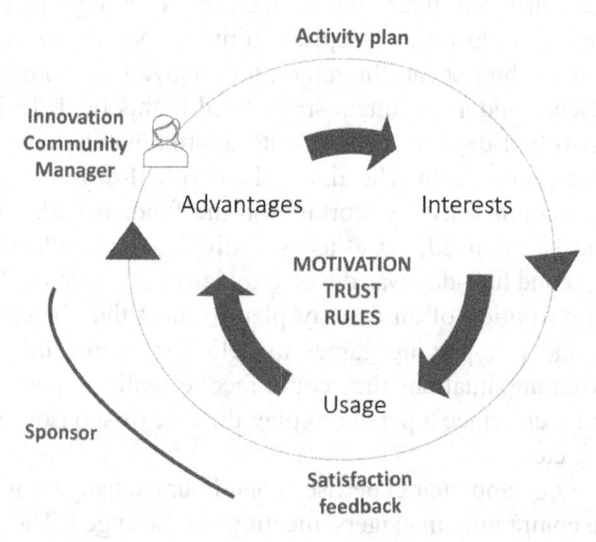

Figure 1: Key factors for the success of a CoP for innovation.

and managing a community. We refer to a *non-managed community* when a group of people organize themselves to share their practice, without necessarily having a dedicated person to ensure that this initiative is managed and that its objectives are achieved.

IV. Key Success Factors Common to Both Managed and Non-Managed Communities

The key success factors described as follows are common to both managed and non-managed communities.

1. *Intrinsic motivation of members: The fuel of the community*

Intrinsic motivation is the overall prerequisite for any form of participation. It is rather an individual factor, which can hardly be controlled, since it is determined by the intrinsic motivation of each individual. If we take the image of a sailboat: without wind … it is difficult for it to move forward. In the case of a CoP, it is the motivation of its members and, consequently, the energy they inject into the life and sails of the community that strongly contributes to its success. Intrinsic motivation is closely linked to the other aforementioned key success factors. First of all, it is related to the community's objectives and theme, which must be of interest to the members. Then, it is related to the practices and the functioning of the community. These may include their modes of interaction: virtual meetings and exchanges, the operating rules adopted, the general atmosphere, the degree of mutual assistance and sharing. And of course, it is fuelled by the benefits derived by each member who contributes and reinforces the overall level of motivation.

2. *A circle of trust*

At the collective level, it is vital to be able to establish a circle of trust between members, so that people are willing to share. At the outset, it is particularly useful to clearly indicate who is part of the community, what every individual's intentions and objectives are, why they participate, and how they wish to become involved. A good way to create trust is to organize regular physical meetings in order to get to know and trust each other, to make the most of the opportunity to arrange individual exchanges (which is not always possible *via* an online platform), to develop affinities and share practices. A second method is to use an online exchange tool, whether it be an email distribution list or a social network such as *What If,* to maintain exchanges between physical meetings.

3. *Working rules that everyone adheres to*

Over time, operating rules will gradually be introduced and will help reassure members and develop routines. It is generally possible to define some community's common values and then define the rituals and functioning of this community. For example: define the frequency of physical meetings (not too often, not too infrequent), the tool to be used to maintain exchanges between meetings, the conditions to be met for a new member to be included (for example, the level of experience expected on a given subject), recording (or not) the minutes of the meeting, *versus* allowing only those who come to the meeting to have access to that information. The implementation and following of these rules contribute to the creation of a relationship of trust between members, which takes time to develop and can sometimes be quickly broken! As the community is a living organism that evolves over time, there is an enduring need to adapt to its objectives, functioning and associated rules, always with the aim of ensuring that the members adhere to these principles and are satisfied.

V. Key Success Factors Specific to Managed Communities

After presenting the key success factors common to both managed and non-managed communities, we now focus on the factors specific to managed communities.

1. *The community manager at the service of member satisfaction*

The specificity of a managed community is that of being led by a community manager, also called a leader or initiative leader. This role can of course be shared by one or more members of the community. This person will be in charge of: co-defining the community's objectives, recruiting members and facilitating their integration, imagining and developing value-creating applications, co-defining and implementing the plan of action, facilitating the integration of new members, relying on members' motivation to stimulate the engagement, and regularly collecting their feedback in terms of satisfaction. This or these person(s) must be leaders, who are recognized and considered to be legitimate within the CoP, who will be able to revive momentum when necessary, and ensure that a good level of contact is maintained between members. In our opinion, it is more the motivation of individual members than the involvement of a facilitator which ensures the success of a community. Nevertheless, this person still greatly facilitates group dynamics, thus acting as a facilitator or even "conductor" promoting collective advancement. He/she oils the machinery and helps to reinforce the aforementioned key success factors.

2. *An activity plan to set the tempo*

This is one of the tasks of the community manager(s), who will have to propose an annual activity plan, in accordance with the objectives to be achieved by the community. The role of the latter is to propose a "macro rhythm" to the community, with highlights that will leave their mark on the members' minds. This could be a learning expedition to London or one of the CoP's key themes, the annual meeting such as the Groupe SEB Innovation Forum, … etc. To this, we can add other events that will provide opportunities to meet and share, such as conferences, the presentation of internal or external experts, participation in a training course, a lunch to raise awareness on a particular theme. Recently, for example, the Groupe SEB launched a conference series on the theme of the Makers' movement, innovation and the Fablab. The important point here is to create a pattern of meetings, accompanied by an ideally intense pace of publications. The most successful communities are always highly topical, where physical meetings fuel virtual exchanges, and vice versa. The combination of both generally leads to strong momentum and membership support.

3. *A culture of permanent feedback*

The role of the community manager is also to continuously collect feedback from members concerning their satisfaction with community participation. This is generally carried out every 6 months in sub-communities. In other companies, the ROE (Return On Engagement) is measured to determine the extent to which members would recommend this community to others. The community manager must do his/her utmost to ensure the functioning, organization and tools that contribute positively to the achievement of the community's objectives. The goal is not to take a long time to do something to perfection. On the contrary, the challenge is to quickly achieve a MVP (Minimum Viable Product) with one's community, as advocated in the book *Lean Startup* (Ries, 2012).

The objective: to progress by iteration. For example, by testing the first possible uses in a functional but basic environment, then gradually adding new functionalities, depending on the need. For example: on the request of the users, add a gallery specifying the responsibilities of each member of the team to bring clarity. It is by testing things in real life that we become aware of real practices and needs.

It is then as the need arises, with feedback loops and a "test & learn" mindset, that the community manager and the members of the community will gradually succeed in finding the well-known virtuous circle. This is why it is useful to ask for feedback during private conversations with members, but also via satisfaction surveys after each physical meeting. The challenge is to be able to identify areas for improvement, and even readjust the community's objectives and practices. The use of an online social network with microblogging and survey features greatly facilitates the dynamics of feedback and participation.

4. *A sponsor who facilitates and gets involved at key moments*

Finally, within the framework of a managed community, the presence of a sponsor who actively participates reinforces the actions of the community manager and the dynamics of the community. The sponsor must be recognized for his/her expertise in the field, must occasionally make presentations, and also encourage and legitimize the existence of the community. This person can also contribute by connecting with other CoPs, thus reinforcing the dynamics. He/she makes it possible to give credibility and legitimacy to the actions that are undertaken, and can sometimes even motivate certain members and their local managers to join in and participate. He/she generally makes it possible to justify the CoP's relevance and necessity, providing a direct contribution to the achievement of the company's strategic objectives or orientations.

VI. Advantages and Evolution of the Community for Innovation

In the following analysis, we outline the main benefits this CoP has yielded for its members and the Groupe SEB, since its creation.

1. *The benefits observed within the CoP for innovation*

The CoP for innovation is truly complementary to the innovation process and the external ecosystem of the Groupe SEB, facilitating transversal exchanges and internal mutual support. In particular, it makes it possible to open up and accelerate the flow of information, by strengthening mutual assistance between its members and sharing knowledge within communities of practice.

The creation of this momentum was possible thanks to some of the community members, who are particularly active and committed. They take part in physical events, share within communities of practice of varying sizes, in real life and via *What If* ... etc. These members make up a core group of approximately 250 people, who we refer to as *intrapreneurs* or initiative leaders, including the 25 community managers. It is their commitment and motivation that brings life to the entire community.

Each member is defined by his/her adherence to the innovation community in the broad sense, and to all the sub-communities of which he/she is a member. For example, a designer could follow several communities: his "Designers" community for the sharing of the latest inspirational trends (colours, materials, etc.) with the Group's other designers, the "proto-models" community for the sharing of questions with experts on the subject, the CAD (Computer Aided Design) user community to learn about the latest features, and the "sustainable innovation" community to be informed of the latest updates on this theme.

The main benefits observed by these people, through their involvement in the community, are as follows:

- Finding a given expertise more quickly (notably thanks to the directory). For example: by typing "UX", one can find all of the people who are specialists in user experience.
- Solving problems more quickly (thanks to the questions asked). For example: by asking if anyone is familiar with a material that is able to

maintain a certain colour and shine, while being heated to a certain temperature.

- Fostering a sense of belonging to a community. It is always enjoyable to feel that one is part of a big family that can be counted on when needed.
- A more effective exchange of information, in particular the monitoring of new developments, by attending physical events and online communities.
- Better sharing of knowledge and practices, in particular within communities of practice.

The CoP for innovation thus offers a "dynamic" network for mutual support and the sharing of knowledge and practices at two levels: that of the community as a whole, and within the sub-communities of practice.

In general, it is not really possible to say whether a particular innovation has emerged through *What If* or the CoP for Innovation. On the other hand, we have observed concrete examples of initiatives instigated through the management and animation of the innovation community. Indeed, it is often a succession of small events that can lead to new initiatives.

For example: a new methodology for collecting consumer insight *via* the web was tested by organizing an online brainstorming session with 80 real consumers for one week. This initiative was created through regular events organized within the CoP for Innovation. The results were highly conclusive: the collection of many insights and *pain points*, suggestions for the improvement of some products, feedback on concept reports, a podium with the three most interesting concepts, etc.

Let's see how this initiative emerged. It all started with the *Innovation tools* community, which set itself the objective of introducing new innovation methods. The community embodied this through its activity plan, by organizing a series of conferences and in particular a *morning session* on *Design Thinking*[1]: offering people interested in this approach to innovation the opportunity to meet over breakfast, to listen and share ideas with several internal and external experts on the subject. During this *morning session*, an internal anthropologist presented a range of topics related to his profession, and mentioned netnography as being increasingly used.

[1]On this theme, we would like to cite the website http://veroniquehillen.com/ where the reader can find more detailed information on *Design Thinking*.

This is a qualitative survey technique that uses the Internet as a data source, and relies on virtual consumer communities. At the next innovation forum, the Consumer & Market Intelligence team announced the imminent launch of their new online platform for consumer insight, which gave rise to the idea of testing a full-scale online forum with 80 real external consumers, using the proposed platform. This topic was proposed by a Marketing and Research team from one of the business units, and coaching with an external company was used to test this new methodology. It is thus the steering of the innovation community and the organization of various events which made it possible to launch this initiative.

2. *The main benefits observed within the sub-communities: The example of the water sub-community*

The more strongly the CoP's theme is directly related to operational issues and its members' objectives, the more motivated they will be to participate, since the advantages they gain from it are directly applicable to their daily work.

We propose to illustrate these through the example of the "Water" CoP of the Groupe SEB. Following an innovation day on the subject, an internal water treatment expert suggested to other colleagues, who were also experts on the topic, the idea of launching this community. The objective: to share on-going studies, including identified problems and solutions. The main benefits observed were the recognition by peers, and the acceleration of their innovation projects.

The first advantage: they could gain access to information that was officially unavailable: either because it was confidential or because it was only "in the heads of the experts", but had not been formalized. Instead of starting from scratch, each person was able to draw, from the topics presented, any elements that were useful for their own project, and explore that topic in greater depth. Thus, by participating in a community, experts could gain access to unique content such as confidential study reports, informal responses on the wall of conversations that otherwise would never have been formalized, due to the experts' time constraints, or feedback from a colleague on a given project. Obtaining information well before it is formalized through traditional channels can lead to a significant competitive advantage.

The second advantage is the mutual support network. Indeed, the interest for CoP participants goes far beyond the ability to draw on information provided by the collaborative tool. Participation makes it possible to socialize and thus to benefit from a mutual support network. It is not a question of finding an off-the-shelf solution, but of developing an original solution together. This shows that the more the members know each other, what others are working on and what on-going problems they have, the more they will be in a position to "provide the right information at the right time" and to help each other. An excellent way to do this is to systematically organize a round-table discussion at the beginning of each meeting, for participants to share their knowledge of current events, as well as their expectations or needs. This makes it easier to connect with those who can help and those who want to be helped. In some communities of practice, this is referred to as the "Post-it ceremony". Each person notes their needs and questions at the time, then presents them orally to the other members. This work then facilitates informal discussions and the collective construction of a solution to the problem encountered.

The third advantage is peer recognition. Indeed, beyond the very concrete and practical motivations to find a solution to a given problem, the glue of community participation is that everyone benefits from recognition from their peers, by showcasing the work that has been completed, and becoming involved in the community in their own way. Receiving constructive feedback and recognition from other experts on a given topic can be a source of gratification. This motivation through peer recognition is particularly useful when recruiting highly skilled experts from communities who do not think that they need to search for solutions beyond their own expertise. Without this intrinsic motivation, there would be a considerable risk of the community declining, due to the desertion of its experts. In addition, the sharing of an individual's background, risks taken, setbacks and achievements, allows him/her to promote his or her expertise with peers, obtain their recognition, and sometimes find the energy needed to continue his or her research and investigations.

VII. Conclusion

The case of the Groupe SEB's CoP for Innovation is an interesting example of the implementation and development of a managed community of practice. In just 5 years, genuine rituals have been established, such as the Innovation Forum, the organization of "on-demand" innovation events, and the regular submission of summary messages to inform members and promote contributors. In addition, many sub-communities have emerged, such as the "sustainable innovation" community. These allow contributors to help each other, share unique content and grow together. Finally, thanks to its internal social network *What If*, the CoP for Innovation has effectively initiated its own digital transition, and gained in maturity in terms of launching and implementing new sub-communities of practice.

As discussed above, the challenge is to succeed in achieving the company's objectives, such as reducing the duration of innovation projects, while at the same time allowing each member to develop and benefit from the solidarity of his/her peers. In our opinion, the success of a CoP depends largely on the motivation and trust of its members, through sharing and helping one another. It is important to successfully establish a virtuous circle, where objectives, practices and benefits are perfectly consistent. The next step for this CoP for Innovation will be to capitalize on the experience gained in the field of communities and digital technology, to expand usage to a larger scale, and to enhance the degree of commitment of its members. The challenge: to imagine new approaches not only for internal collaboration, but also for interactions with external partners, with the aim of deriving maximum value from each innovative ecosystem.

References

Arzumanyan, L. and Mayrhofer, U. (2016). L'adoption des outils numériques dans les communautés de pratique: Le cas du Groupe SEB, *Revue française de gestion*, Vol. 42, No. 254, 147–162.

Ries, E. (2012). *The Lean Startup: How Today's Entrepreneurs Use Continuous Innovation to Create Radically Successful Businesses*, Pearson.

© 2021 World Scientific Publishing Company
https://doi.org/10.1142/9789811234286_0006

Chapter 6

The Schmidt Groupe Organizes Ideation with the Créativ'Café

Tristan Cenier and Patrick Llerena

The firm Schmidt Groupe is probably one of the most innovative in its business sector and has been over a long period. As with many family ETIs [French classification for medium-sized companies], it has driven its development with a long-term vision of its durability and it has known how to anticipate both technological developments in the processes of production, products and the uses of its products. For several years it has established a relatively structured process of innovation, with an organized sequence of precise and selective activities and a *stage-gate* process. However, two observations have been made: on the one hand, actors acting during the innovation process have little opportunity to interact apart from more or less formal meetings organized by the process, particularly in advance of this and, on the other hand, the sources of "inspiration" of the process, which supply it with new projects, are "monopolized" by a limited number of people belonging to "authorized" services to contribute to this process. The question posed is, then, not only that of the existence of an "inventor", of their identity and intrinsic "genius", but also that of their "localization" in the organization and, above all, sharing their idea with a group, a collective, to enable the idea to mature and be acceptable as a candidate to become a project in the innovation process.

The experiment that we will analyze in this chapter, the creation of a "Créativ'Café" system, will aim to organize the process of ideation by giving it visibility, by making it public and open. This process of ideation is organized in parallel and independently of the process of innovation and is designed to be one of the potential sources supplying the process of innovation in new projects. It will also eventually be the opportunity to reveal the existence of one or several latent communities of innovation within the firm.

In fact, to use the work of Cohendet and Simon (2015) and their representation of the duality of the processes of ideation and innovation, the "Créativ'Café" is a particular form of an ideation process, while *stage-gate* projects represent the firm's innovation process (Figure 1).

These two processes coexist in parallel and probably with their own logic and horizon. The "Créativ'Café" is *de facto* a particular form of the ideation process. The nature of ideation processes should therefore show us the principles of the operation and concept of the "Créativ'Café", as well as its management style. It is too early to draw all the conclusions of this experiment. But its launch and the first cycle of its implementation are already helping us to learn some interesting lessons, in particular on its relevance.

After presenting the context of the "Créativ'Café" experiment, we will tell the story, progress and results of this as a particularly representative illustration of establishing a plan to manage emerging ideas and their integration into what is, on the whole, quite a classic innovation policy.

I. A Brief History of the Schmidt Groupe

The Schmidt Groupe is a family firm, founded in 1934, today managed by the granddaughter of the founder. Initially Cuisines Schmidt, the firm became Salm (Société alsacienne de meubles) in 1983 and distributed its brands, Cuisines Schmidt, then Cuisinella, from 1992, *via* a chain of franchized shops. Five factories (soon six) at four sites made it the leading kitchen manufacturer in France from 2008. Taking advantage of its economic strength, Salm developed agile management, invested in cutting-edge technology, installed new production methods, built model factories of factory 4.0 and relied on innovation. In 2016, to improve its visibility and honour its culture of innovation and its roots, Salm became the Schmidt Groupe.

The Schmidt Groupe operates on a bespoke basis and guarantees a lead time of five weeks between order and delivery of a fully equipped kitchen. "Custom-made" is the core principle of the Schmidt Groupe's offer: the kitchen is designed in-house by the vendor and the buyer, taking account of the wishes and profile of the latter, and the features of the space

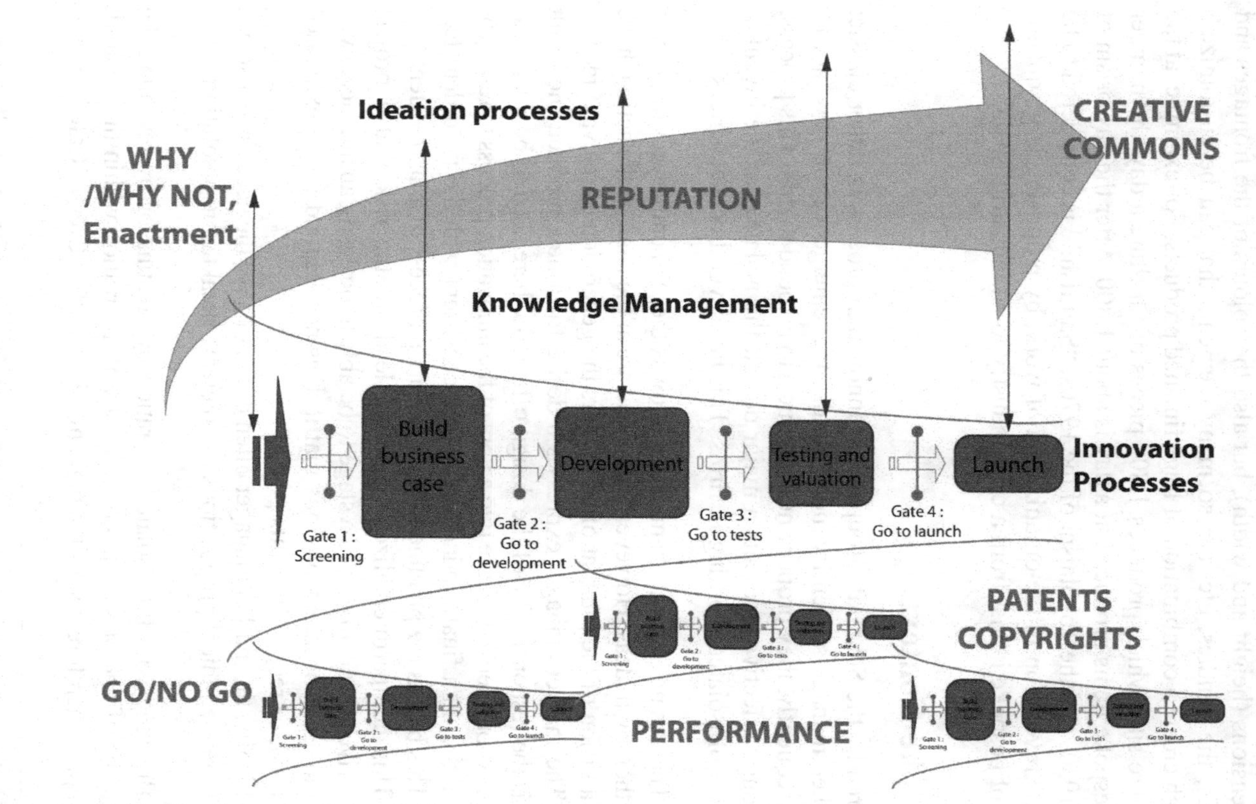

Figure 1: Inspired by Cohendet, Simon, 2015 and presentations at the Autumn Schools of Creativity Management.

in which it will be installed. Ultimately: the layout of the drawer units, arrangement of the units, (drawers, cupboards etc.), as well as their dimensions (height and width), but also the colours of the frontages and edges, the handles, etc. are so many elements that can be customized, which create a combination of huge finished products. For example, at the Lièpvre site, which produces 1,800 pieces of furniture a day, the number of bespoke items produced in a day is about 1,800. After the Taylorism of the 1930s and the Toyotaism of the 1970s, "Salmism"[1] is considered to be a new revolution in the organization of work, by proposing large quantities of bespoke items: from a craft industry to an industrial scale.

II. The Context

Currently, the Schmidt Groupe (SG) innovation process is rather classic: market managers identify new trends, the emergence of new needs. Consequently, they propose new products to respond to this. This process, although effective, presents a major defect: the isolation of the inventor, typical of isolated working in firms. This has several consequences:

1. There is no internal channel that allows other SG employees to express their own ideas. However these, who are at the heart of the action, have a vision of the job that often favours the generation of relevant ideas.
2. The market manager expresses a desire for a new product; the result is the responsibility of the design office. So there need to be many discussions between these two actors of the innovation process to result (or not…) in a final product or service, consistent with the need detected. This blocks the pipeline of the innovation process to a large extent.
3. This compartmentalized operation leads to the risk that a project arrives at too advanced a stage to be able to be easily abandoned, even though it becomes obvious that it does not respond to the expressed need, or responds to this poorly.
4. Finally, the market manager often does not have the necessary distance to make radical propositions — to create breakthrough solutions.

Breaking with this isolated operation by creating or by revealing the existence of one or several communities of innovation within the Schmidt Groupe is therefore one of the main missions of the "Créativ'Café" system.

[1]Neologism inspired by the firm's historical name, Salm, and referring to Fordism, then to Toyotaism as principles for the organization of production and management.

Given its objective, the "Créativ'Café" should eventually be directed at all the employees of the firm. As an experiment, it was decided first to limit it to a small population in order to test the system. At the end of this experimental phase, the target population would be expanded to all employees. The starting objective was to be cross-sectional, it was therefore made up of employees belonging to several services, closely or not so closely associated with the process of product development: procurement, pricing, marketing, design office, design, foremen/forewomen, etc. The community of 90 people was invited to the launch of the initiative in January 2016; almost three-quarters of the participants were present on that day. The director of product development presented to them the principle of the "Créativ'Café", its objectives and its operation.

III. The Principles of the Créativ'Café

- A good idea is the result of collective, and not individual, reflection.
- A good idea is put forward by someone who wants to see it succeed.
- A good idea does not respond to formal obligations.
- A good idea does not conflict with everyday concerns, nor is it limited by them.
- A good idea is the start of a pathway, not an expected outcome. These principles lead to many consequences and questions.

1. *Lack of an obligation to achieve a result*

It is not possible to predict what the result of this initiative will be, not even to guarantee that there will be a result: after several "cafés" on a problem, a subject, an idea, there may be no usable result. In contrast, what is guaranteed is that energy will have been expended! Even if a community can be mobilized on a subject without the pressure of a result, this raises other difficult questions. In the absence of an objective of results, how do we decide how much time an idea should be worked on before abandoning it or considering it to be fully developed? How can we be satisfied with a non-result? And especially, when should a "Créativ'Café" be ended?

2. *Absence of an a priori agenda*

In addition, it is not possible *a priori* to impose a timetable for action. The members of the community voluntarily take time from their work and/or their lunch break to participate in the "Créativ'Café" sessions. They also have many time commitments. It is almost impossible to predict in advance the number of sessions that will be necessary to complete a reflective process. In plain English, we know when a theme starts, we never know exactly when it will finish, nor how it will evolve, nor what the result will be. It is difficult to gain acceptance that here it's a question of an intrinsic characteristic in the ideation process itself ... particularly through the hierarchical structure of people involved.

3. *Two key figures: The sponsor and the facilitator*

The sponsor of the idea is the person who launches the process; who takes the initiative to set out their idea and to share this at an initial "Créativ'café". The facilitator ensures support for the sponsor: s/he advises him/her on the format of the idea and its promotion. S/he is also the resource person to propose suitable methods of creativity, depending on the nature of the ideas, and who follows this up and ensures capitalization from one session to another.

IV. The Objectives of the Créativ'Café

We have already mentioned the first objective of the "Créativ'Café": the production of ideas for new products and new services. There is a second objective, at least as important, if more vague: to change work habits in the firm, introduce change, a disruptive element.

1. *Production of ideas*

The area of work in the "Créativ'Café" is, in three words: "*What, for whom?*".

The *what* is the invention, the product/service that will be proposed. At the end of the "Créativ'Café" sessions, the sponsor should be able to convincingly present the idea developed. To do this, a few sessions will be devoted to producing presentation materials in the group, generally in the form of a model, in paper, in cardboard, in Lego, or any other appropriate material. At this stage it is not about thinking about a technical solution: no prototype is made. The investment in time and resources should be reduced to the minimum. If the idea is abandoned, it should have cost the firm almost nothing.

The *for whom* is the definition of the target, the public concerned by the invention, the end user of the new product, or the beneficiary of the new service. The group therefore works to define the assumed and approximate value of the idea for the user, estimates the size of the market that it will impact (if possible), the advantages compared with solutions that already exist, the benefit to the firm's image, etc. Here as well, at the end of the sessions, the sponsor should be able to present these in a convincing way.

2. *New methods of working*

This is a welcome side effect of the "Créativ'Café" rather than an objective in itself. Showing participants that it's possible to work differently. For example: a one-hour session allows huge progress to be made on a subject, provided that the meeting is structured according to a method adapted to the desired objective. The role of the "Créativ'Café" is also to offer an intellectual space which contrasts with the routine of daily work. For one hour, participants discuss subjects outside their role, expressing opinions, putting forward an innovative project, sometimes allowing themselves to launch crazy ideas that will be welcomed. The "Créativ'Café" is also the opportunity to collaborate with people who you don't often, or never, see. In short, the "Créativ'Café" is a workspace where the work should be "fun", a change compared with the usual routines (even if it is destined to itself become a particular routine).

V. Operation of the Créativ'Café

The practical operating principle of the "Créativ'Café" is based on the following elements:

- The "Créativ'Café" is a space for discussion and experimentation;
- The "Créativ'Café" helps to confront, to challenge one's inspiration with a voluntary and kind group;
- The "Créativ'Café" has only one demand: to help to combine the desire to share one's idea and the desire to contribute to the emergence of an idea;
- The "Créativ'Café" uses few resources;
- The "Créativ'Café" adopts the motto dear to companies in the Silicon Valley: "fail often but fail fast".

Any member of the community with an idea for a new product contacts the facilitator of the "Créativ'Café" and agrees a date for a first meeting with him/her. The facilitator announces the chosen date to the whole community and briefly presents the sponsor's bright idea. The presentation is as neutral as possible (to avoid a *framing*[2] effect as much as possible), while producing the desire to participate. Participants who are interested in the idea have until the day before the "Créativ'Café" to sign up for this. The sessions last one hour and take place during the lunch break or in the evening after work hours. At the end of the session the facilitator decides, in agreement with all the participants, on the date of the next session, thus giving each person the opportunity to track the development of the idea. Once the date of a new session has been decided, the invitation is always sent to the whole of the community. In this way, not only is a hard core formed around the idea and/or its sponsor, but also any other interested person can be added to the group at any time. This behaviour is even encouraged because it is the opportunity to inject a new viewpoint into the reflection, to reveal a problem forgotten by the group, even to build on the discussion and launch the idea in a new direction!

In the same way, any participant can leave the group, for one session, two, or permanently, in agreement with the principle that pleasure is the only justification for one's participation in "Créativ'Café".

The material provided for the "Créativ'Café" is deliberately frugal: paper, cardboard, something with which to cut and glue, something with which to write: a white board, a flipchart, etc. For purposes of simplicity as much as demarcation compared to regular meetings, computers are banned: no PowerPoint!!

There is generally a gap of 10 to 15 days between sessions: less than 10 days there is an overload of work, more than 15 the memory fades.

The number of sessions needed is left up to the group. In fact, at the end of a session, the decision is taken to stop or to continue the work. The process comes to an end when the idea has been sufficiently worked on to become a "good idea", namely, an idea for which there is a credible response to the questions "What?" and "For whom?".

The formal conclusion of the "Créativ'Café" is the presentation of the ideas developed to a decision-making committee. This committee consists of marketing managers and managers from several departments linked to

---

[2]Creating an intellectual framework that will limit the imagination and boldness of the participants.

product development. For each project, the committee's decision can approve the idea, which will then join the pipeline of product development (innovation process), a request to rework the idea in the "Créativ'Café", or a rejection of the idea. Of course, the latter decision can be difficult to accept for those who have developed this, and especially for its sponsor. Failure is therefore an integral part of the process and the members of the group and, in particular, the sponsor, must be convinced of this. It should therefore not be perceived as a negative conclusion. On the contrary: ending one subject helps to free up time to develop other ones. The sponsors, whether their ideas are rejected or not, are (should be) recognized as driving forces of the community of innovation and are strongly encouraged to put forward new ideas and/or to take part in other ongoing "Créativ'Cafés".

VI. The History of the Créativ'Café

The "Créativ'Café" initiative was announced to members of the community during a launch meeting. Immediately afterwards, members of the community who had an individual idea for a new product or service and who wanted to develop this within the firm were invited to contact the facilitator to schedule the launch of their "Créativ'Café".

The first campaign of the "Créativ'Café" — namely between the call for individual ideas and the first selection committee — took place over five months. Six ideas were developed, support was rapid: sponsors came forward in the first week after the launch. These six ideas were very varied and, for this first campaign, were only about products (no services), which was consistent with the existing culture in the firm.

1. *Storage makes the kitchen*

Proposes kitchen furniture be built on the same principle as storage furniture. Since 2004, the Schmidt brand has in fact offered storage furniture designed as a large bespoke unit, divided into several compartments/functions (wardrobe, bookcase, shoe rack …), flat-packed and assembled on site by the fitter. As for the kitchen offer, this is mainly based on a collection of drawer units. The idea here is to develop a kitchen line based on a single large piece of furniture, uniting all the desired functions *via* subdivisions of the furniture space.

2. *Origami*

After the *2-second* Decathlon *tent*, here's the Schmidt Groupe *2-second kitchen*, furniture which unfolds, ready to install. The Schmidt Groupe offers a range of furniture in kit form, which struggles to find its place faced with the Swedish giant which already dominates this sector. This idea proposes renewing the furniture in kit form: no longer any need for tools or for instructions, a reduction in damage, so many elements that can improve the experience of the user.

3. *The variable geometry kitchen*

A kitchen of variable dimensions in the space, which is adapted to the needs of the user. A large kitchen helps to organize meals for large groups: a family lunch, cocktails, etc., but it is often also a luxury of space. In contrast, a small kitchen helps to usefully save space, especially for those who have most of their meals outside the home. But why choose? Why not have a kitchen which is both small and large? This, very fundamental, idea aims to resolve this contradiction by providing variability in the spatial layout of the kitchen.

4. *Architectural finishes*

Let's break the aesthetic uniformity of the fitted kitchen. A fitted kitchen, as it is a compact unit of parallelopipedons, is presented, among other things, as a group of horizontal and vertical lines. This Café's starting idea is the creation of an aesthetic line based on decorative strips, which helps to break this overly systematic aspect. The starting concept has gradually evolved and the subject has become a major change to the very layout of the kitchen.

5. *Functionality and aesthetics*

A kitchen should be practical to use and it should be nice to look at, all the rest is superfluous. This is an idea that is a complete break from the current professional knowledge at the Schmidt Groupe. The plan here is to get rid of the concept of the drawer unit: the kitchen as a whole is based on a light framework that enables hardware items (door hinges, drawer slides, etc.) to be hung, to put on the worktop, and decoration using thin panels. As these no longer act as structural support, they could be changed regularly, depending on how the owner's taste changes.

6. *The kitchen from 7 to 77*

The kitchen is a space which should be accessible to all users, whatever their age. In accordance with this principle, the idea of this "Café" is a system that allows a child to reach things that are put away high up without climbing on a chair, just as much as for an elderly person to reach things put away low down without having to bend down. This offer should be made at a constant price, which excludes any use of a motorized system. One solution proposed is a system of rotating trays, meaning that the object the user is looking for is always available in his/her comfort zone.

The population of the participants in the "Créativ'Café" for all subjects consists of a majority of men (two-thirds), 34% are older than 45 and 24% are under 30, half have been in the firm for 10 years or more. Compared to the community that was asked at the start, they represent just over a third (39%), or a total of 35 people. Asked *via* a questionnaire, 38% of participants considered themselves to be creative, while 50% were undecided on this question. However, 12% claimed to be "not at all creative": these are the brave ones because, despite this, they participated in a "creativity" initiative. The sub-population of subject sponsors followed the same demography exactly, with the exception that they mainly see themselves as creative people.

A group formed around each of the themes described above, the number and composition of which has generally remained variable (Figure 2).

The average number of participants was 4.36 ± 1,8 per session, for all subjects. We note, for the future, that with six themes and five sessions on average per theme, the data sample is too low to achieve statistical significance. However, we note that the standard deviation is not correlated to the size of the group. The change in the number of participants is also not one way: some groups were reduced to two people for one session, returning to their normal size in the next one, sometimes even with new arrivals. In contrast, we have observed a hard-core phenomenon for all the groups. The sponsor is always present, by definition, but there have always been one or two people who are loyal, sometimes more, who

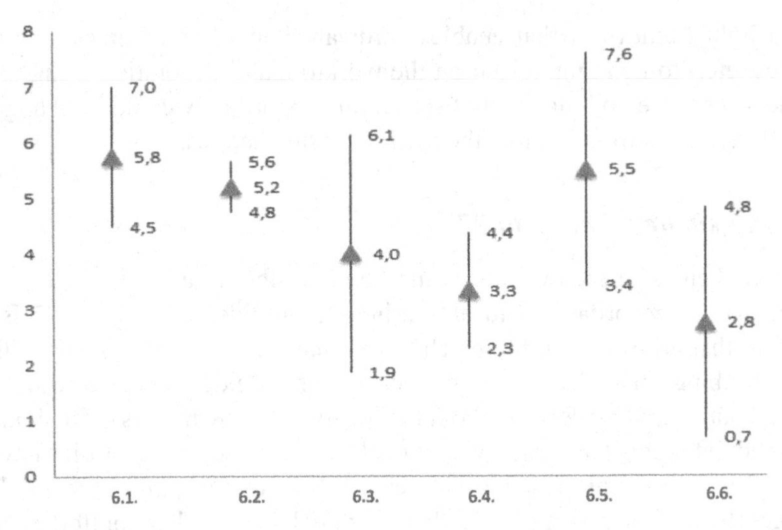

Figure 2: Average size of groups calculated for all sessions on the same subject in the number of participants, more or less one standard deviation.

followed the sessions to the end. The average rate of attendance is 65% (namely, that a participant attends on average 65% of sessions on a subject, or several subjects). A majority of participants (66%) only attended on one subject, with these the attendance rate was 70%, so above the average. Conversely, with those who participated on several subjects (two subjects: 21%, three and more: 13%) the attendance rate was not more than 55%, so some people flit from one subject to another, so one must be vigilant when faced with the possible emergence of tourists, understand how to make good use of them, or study the motivations of their behaviour: a lack of information equal to their curiosity, a fear of, or refusal to, commit … It should be noted that, among subject sponsors, 83% of them have participated in at least one other subject as well as their own, with an attendance rate of 30%. With the sponsors, flitting about cannot logically be attributed to a refusal to commit, so perhaps one can see in this an excess of curiosity?

In general, a dual phenomenon of volatility and crystallization emerges from this dynamic of the population of participants.

After having consulted the members in question, *volatility* is largely explained by professional constraints (business trips, meetings, etc.), but also by a loss of interest in the subject. This behaviour was expected because the Café operates on a voluntary principle: one only participates in a subject as long as one wants to do so; it is, even, reassuring to see that this principle has been well accepted.

As for *crystallization*, this is quite simply explained by support, both for the project but also for the "Créativ'Café" as such. One participant demonstrated this in this way:

Concerning my motivation to return to the "Créativ'Café", there is the human aspect, the life of the group, the quality of the subject and the unknown, this characteristic to be created, to find a solution to the subject. This short hour also helps to relax, a valve where there is steam, energy to be recovered, it's a way of recycling which can be productive at all levels.

It appears, particularly *via* crystallization and the relatively low *turnover* of the groups, that a community is emerging around the "Créativ'Café" and that it is producing a certain support. A more careful analysis over time will be necessary to know if this is about an effect linked to the novelty of the exercise or a structuring phenomenon in the organization.

VII. Roll-out of the Session

A session of the "Créativ'Café" takes place over an hour, generally from 12.30 to 13. 30. The participants voluntarily take time from their lunch break to take part. This slight obligation helps to guarantee the motivation of the people who are present. There is therefore a "cost" to participation.

Processing of a theme is organised into three segments, spread over the total number of sessions: 1-definition of the problem, 2-definition of the *what,* 3-definition of the *for whom*.

- Definition of the problem is the most crucial stage, it allows participants to take note of the detail of the sponsor's idea; it sets the outlines of the idea which will emerge *in fine* and it also helps to outline the one or several sources of value for the firm and for customers. The approach followed is to consider the idea as a response to a problem, to an issue; it's about identifying it. For example, the idea "The kitchen from 7 to 77" was that of a rotating mechanism carrying a set of shelves, fitted in a drawer and allowing the contents of this drawer to be picked up without bending down. At the first session of this

Café, the participants thought about the organisation of the kitchen layout, analysed that the storage space most used for food and the most common items is located at the level of the hips, and concluded that the problem was:

"What do we do to make sure that all the storage in the kitchen is located at the level of the worktop?". The starting idea was a response to this question but, from that, the group could have identified and developed other possible responses and presented another idea to the decision-making committee.

- The definition of *what* is certainly the most "natural" stage to implement. Above all, one must make sure not to start this before all the participants have an exact idea of what has to be done. Once the stage of defining the problem has been concluded, the response to the "what" is already there and focused in each person's mind. Now it's about representing the idea by using a "model", a suitable mechanism to make it understandable to actors outside the group, clearly and practically; and in particular to the decision-making committee. Consequently, this model serves neither to present a product in its final end state, nor to present a technical solution. It doesn't even have to be "beautiful", it's only useful as an illustration. Instead of a model, this is a 3D sketch. The material chosen for this is cardboard, but other media or materials can be used: for example building kits (Lego, Meccano…), even 3D printing, if this is justified.
- Final segment, defining *for whom* is the stage that has posed the most problems in this first campaign. The main reason for this is that participants in the group, including and, above all, the sponsor, are very enthusiastic about their idea and naturally believe that the end product will be of interest to all potential customers. It's the responsibility of the facilitator to steer them to understand the target of the idea in a more precise way. This stage of defining *for whom* occurs at the end of the "Créativ'Café". The group then evaluates the idea that has been developed: its principle, to what it is responding, what is new about it, possible objections to its implementation, benefits for the end user, etc. However, the exact identity of this end user was always carefully avoided by all the participants. During this first campaign, the facilitator was not able to convince of the importance of this aspect. This will be a significant point for improvement in future campaigns.

VIII. Problems Encountered and Initial Thoughts After a First Campaign

1. *Acceptance of the Créativ'Café*

The "Créativ'Café" is not like any usual practice in the firm: in particular, the principle of the lack of obligation of a result. This position is an advantage for the initiative because it attracts the curious, the "alternatives", for whom enjoyment is an essential component of work. But this can also pose a divisive problem of legitimacy; a tarnished image with those for whom efficiency and the result take precedence. Therefore, the first key to success of the "Créativ'Café" will be to accept that it has to prove itself.

- It's not because there is no obligation to arrive at a result that there is no obligation to set objectives: the aim of the "Créativ'Café" is to come up with carefully thought through ideas: if five subjects are being developed, the objective is to present five of these. Therefore, this objective must be defined correctly very early on and the firm's internal communication should be committed to pointing this out.
- All those who have already taken part in a session of creativity know this: the experience is enjoyable, sometimes even exhilarating, but above all exhausting because the work and the attention are intense. To help the credibility of the "Créativ'Café", especially at the beginning, particular attention should be paid to this aspect of intense work, by explaining that this is a different kind of work. It is probably better to place the spotlight less on fun and games — no doubt a French cultural particularity, but which should be taken into account. The experience, results and "word of mouth" are sufficient to increase the attractiveness of the experience.
- It should also be considered that the "Créativ'Café" is not the first attempt to develop creativity or innovation in the firm. The "Créativ'Café" can be seen by some colleagues as "yet another new creativity thing with no future". The major disadvantage with this flawed legacy is to divert colleagues from the initiative who are fully in favour of it but who don't want to waste their time. There as well, demonstration by results and their visibility and impact are and will be the only antidotes.

The length of time is probably the most important variable for the success of the "Créativ'Café". This duration allows it to prove itself and to establish routines — to bring operating methods into the firm's traditions: the lack of an obligation to produce a result, investment in staff time, autonomy and freedom of action.

The "Créativ'Café" practice is built on a "snowball" effect, the first participants will relate their experiences and will encourage new people to engage. The community of pioneers are guarantors of the future development of the community.

So we find ourselves with two parallel processes, of different kinds: the ideation process of the "Créativ'Café" and the innovation process, which is based on a conventional "*stage*-gate" approach, oriented to deadlines (and not length of time) and result.

2. *Organization of the sessions*

The organization of the "Créativ'Café's schedule is also an important key to the proper functioning of the initiative. This concerns both the roll-out of sessions and how they are spaced out over time. Regarding the organization of time, the main difficulty is adding to diaries that are already very full and very different, often comprising work meetings that clash with the "Créativ'Café's" timetable. But, ideally, the sessions should be fairly close in time: if the frequency is too low, forgetfulness leads to repetition and to a fall in motivation. Therefore, a good compromise must be found between the time that each participant can devote to the "Créativ'Café" and the necessary minimum frequency of sessions.

For the roll-out of the sessions, we must consider three time-consuming hindrances: updates, delays and purges.

Updates: at the start of each session, a few minutes are devoted to recalling previous instalments, this is all the more necessary because each session can welcome new participants. The shorter the period of time between two sessions, the shorter, even unnecessary, are the updates, with longer-serving participants able to quickly bring new ones up to speed.

Delays: with one-hour sessions and where the end time can't, in principle, be postponed, a delay of 5 or 10 minutes significantly reduces the time available.

The purge: a known phenomenon of regulars at creativity sessions, the purge is the first phase when participants remove common ideas, down-to-earth or fanciful, a stage that is difficult to control, that can't be reduced but which is necessary before beginning to produce innovative ideas. In the context of the "Créativ'Café", participants come to the sessions after a morning at work, they are thus immersed in their daily life and consequently particularly inclined to a long phase of purge. The short format of the "Créativ'Café" therefore becomes a disadvantage: if the purge lasts 20 minutes the usable time will be reduced considerably, especially so when it is added to delays and a long update.

To get round these obstacles, the organization of the sessions should be structured and the "Créativ'Café" has selected some rules:

- At the end of the first session, plan the dates for all the next sessions, in agreement with the participants. This introduces some rigour into the organization of the "Créativ'Café", but allows avoidance of schedules that are too full within a two-week period. The rest of the "Créativ'Café" community is informed of the dates once these have been decided. It is easier to cancel a session that is not necessary than to add one urgently.
- Before the "Créativ'Café", set aside time for a welcome period: a comfortable and informal time, helping to both absorb the delays and to carry out the update in a smooth way. In this way the reflection begins gradually and this creates a "decompression chamber" between operational activities and the creativity session. This welcome period can take the form of a lunch, for example, or it can be done around the coffee machine.

3. *Putting the cart before the horse*

Even if one must not be prevented from being ambitious, care should be taken to remain within the area of expertise of the "Créativ'Café". More precisely, two supplementary pitfalls should be avoided: launching into seeking a final technical solution and restricting the project, self-censuring in some way, for fear of the technical impossibility of producing the invented product. In other words: one must not allow oneself to be overwhelmed by technology… In order to navigate between these pitfalls, the experiment proceeded in the following manner:

- Limiting oneself to the resources: to function, the "Créativ'Café" has simple building materials (paper, cardboard, scissors, glue). The "inadequacy" of the resources allows one to focus on the idea. For example, one of the subjects was a piece of furniture equipped with rotating trays to improve its ergonomics. Only having cardboard, it was of course out of the question to make cog wheels which allowed operation of the rotating mechanism; fine! since the aim of the "Créativ'café" was mainly to have an overall idea of the original function proposed by this invention and to reflect on its target audience. The invention and development of the mechanism itself represent a later phase of the process of innovation. By limiting us to the resources we have, we stayed within the area of expertise of the "Créativ'café".

- Focusing on the value of the idea and the innovation which came from this. We must pass on the following message very clearly and from the start: a determined sponsor will always find the necessary resources to implement a well-defined innovative idea, with strong economic potential, direct or indirect.

IX. "Don't Overdo It ..." or How to Stop?

One of the difficulties (*de facto,* poorly anticipated) of organizing the "Créativ'café" is that of ending the subject... And resolving this is decisive for the viability and therefore the duration of the experiment as a whole... in the interest of revitalizing campaigns.

The "Créativ'café" is above all a platform for shaping and validating ideas that emerge in a community (-ies) of innovation in the firm. One of the important modalities is that of the rapid rejection of ideas not leading to a potentially interesting innovation for the firm: "Fail often but fail fast". Ideally, it should be the work group itself, or the sponsor of the idea, who decides on when to stop. But this proves to be difficult: the group dynamic does not include its own disappearance for two reasons: first, attachment to the subject itself, then attachment to the slogan "Créativ'café" after that.

The driving force of the "Créativ'café" is an intrinsic motivation for participants and especially the person who proposes a personal idea. After working on the subject over several sessions, the interest becomes attachment. It is therefore difficult to accept, individually or collectively, that the process will not end in a potentially interesting idea or innovation. This is also true to a certain extent of the "Créativ'café" itself as a social space: united in effort, the members of the group like each other, as they like this break in their regular work. Separation or renouncing this space, at least as much as abandoning the idea, can't be left to the sole initiative of the group. External regulation is necessary to shift the responsibility towards the organization and governance.

It very quickly becomes evident that it should not be the responsibility of the facilitator to declare a "Créativ'Café'" "pointless". This would make his/her role and her/his position in the dynamic of "Créativ'café" ambiguous and would therefore damage their task of facilitator and contact person during the process. As an active member of the group, s/he naturally takes part in evaluating the progress of the discussions and the idea, but without having a position of "authority".

It quickly appeared clear that the decision to halt a subject was not in the area of expertise of the "Créativ'café" itself. The possible ending will therefore be the task of an external expert assessment which decides on the future of the idea developed. In our case, this expert assessment took the form of a "decision-making committee". However, it is obvious that as participation in a "Créativ'café" is voluntary, it could also end naturally through a lack of participants or the explicit decision of the participants. This was the case with one of the ideas of the first campaign.

X. And … end of a First Cycle or how to Conclude a First Campaign?

1. *The decision-making committee*

A decision-making committee was established to define the future of the six ideas considered as part of a "Créativ'café". This consists of half a dozen managers responsible for services involved at different stages of product development in the firm. It has managerial (decision-makers relevant to innovation) and technical credibility, with people who are able to evaluate both the technical and economic relevance of the idea proposed.

2. *Mode of presentation (pitch)*

To echo the desire of the "Créativ'café" to set up new working methods, the presentation of ideas to the decision-making committee occurs in a new and effective way in the firm: that of the *elevator-pitch*: a total of 10 minutes per idea, the sponsors present their ideas in two to three minutes, then the committee has five minutes to ask precise questions, then two minutes to fill in an individual assessment form. All the sponsors thus appear one after the other, according to an order decided on in advance. This only takes a very small amount of time in the day and can be set up very quickly. The committee then has to discuss and make its decision on each project.

The sponsor uses no IT materials in her/his presentation. S/he tries to clearly present the two parts of the diptych: "What" and "For whom". S/he bases his/her talk on the model made during the Créativ'café" to present the "What" and has written materials (white board …) if required for the "Who".

3. *Assessment criteria and possible issues*

Members of the committee fill in an assessment form during the sponsor's presentation. Very simply, this form has three criteria:

- *Innovation*: Innovative nature, degree of rupture compared with existing products.
- *Adoption*: Size of the market targeted by innovation and facilitated by adoption.
- *Appeal*: Attraction to the project, probably the most important criterion at this stage of development.

Each criterion is noted individually, the total score serves to support the decision.

Three different decisions can be taken by the committee:

- *Rejection*: Rejection of a project is in the DNA of the "Créativ'café". The principle "Fail often, but fail fast" then becomes effective. Rejection of an idea is not a failure but instead a stage that is expected in any creative process. Ending the development of an idea which does not appear to be promising helps to free up resources to begin the development of other ideas. Sponsors of an idea that has been rejected are expressly invited to present new ideas in the "Créativ'café".
- *Continue*: The idea presented is rejected, however, it emanates from a very broad or very vague, but promising, idea. Sponsors are then invited to continue to work in the "Créativ'café".

- *Approved*: The idea presented is adopted by the committee: either it is judged favourably by all the members, or one of the members has strongly expressed her/his support (a "joker"). An approved idea becomes a project, which will then be injected into the standard pipeline of the development process of new products with the Schmidt Groupe, with an important particularity: the sponsor remains attached to the project. S/he will play an active role in its development and in principle could support it until it is realized.

4. *Results of the first campaign…*

Out of the six initial ideas, five made it all the way to the end of the process.

The "abandoned" idea had a particular evolution: the brainstorm produced an idea that the group spontaneously abandoned. To date it is still difficult to judge if the idea was good or not, the fact remains that, session after session, the group did not succeed in giving it shape, nor to provide a clearly modelled example, nor to determine what would be the target audience, so there was no "What", nor "For whom". The role of the facilitator was to relaunch a session to analyze the problem, to create a new trajectory. Finally, although the idea basically remained the same, the group has since worked on a new, much more promising, interpretation.

The five remaining ideas were presented to the decision-making committee. Two of them were rejected.

The committee decided that the costs of research and of modifying the production line and production methods were not justified by the potential commercial value of the end product. In both cases, the quality of the idea was not questioned, it was the weakness of the commercial prospects and the low degree of innovativeness that were the reasons for the decision.

Two other ideas were approved.

For one of these it was its radically innovative character that appealed to the committee. However, this involved a profound reworking of current production methods and the acquisition of new job-related skills. However, it presages a possible evolution in what the fully fitted kitchen will be in 20 years' time and this long-term vision is priceless. For the second idea selected, it was its original character and its air of "curiosity" which appealed. It is still difficult to say to whom this new product will be addressed (the "for whom"?), and if it will be accepted, including internally. However, on the one hand, it is sufficiently appealing to make people want to continue with its development and, on the other hand, this product will contribute a lot in terms of image for the Schmidt Groupe, showing that the firm knows how to take risks and propose original products.

It should be noted that the committee as a whole was in favour of these two ideas. Furthermore, each of the ideas was adopted by a member of the decision-making committee. These members acknowledge that they were particularly attracted by the idea and passionately wanted to see the project succeed. These two people therefore placed themselves in the role of sponsor, without a direct role in the development, but exerting a beneficial influence.

One idea was finally judged to be too timid compared to the starting idea which, on the contrary, was very promising; the sponsor was quickly encouraged to continue the work in the "Créativ'Café".

Final result: the process begun and tried by the "Créativ'Café" has convinced the decision-making committee and its continuation and development have been approved. This approval was achieved through recruitment for the post of facilitator of the "Créativ'Café" and, more broadly, the ideation process. The capacity of the scheme to mobilize creative energies in the organization has been recognized, with a transformation of individual intrinsic motivations into a collective dynamic within the "Créativ'Café".

XI. Conclusion

The experiment of the "Créativ'Café" is a particular method of development and the management of ideas to become projects; a complementary and parallel process to the innovation process in the *stage-gate* process, which also exists in the firm.

These two processes are complementary and are mutually reinforcing. There are often people who are involved in the two processes. More precisely, the sponsors of the "Créativ'Café" are also active in the innovation process. They reveal their need to find new areas to express their engagement in developing creativity, innovation and more broadly the success of the firm. Their involvement in these "Créativ'Cafés" is symptomatic and indicative of the difference in nature between the two parallel processes. There are clear indications that, for the ideation organized by the "Créativ'Café", we have a strong intrinsic motivation, significant autonomy of the participants, and an *a priori* lack of obligation of results. However, it is necessary to ensure management through the leadership of the sponsor, whose personal appeal (reputation) is probably just as decisive as the proposed idea. The "Créativ'Café" is a space where the usual rules of the organization are explicitly waived and where new forms of work are used: this is *de facto* an internal space of freedom. The experiment shows that there is a community (or more likely several) that is interested in occupying this space and to take it over. Campaigns to come will help to confirm the durability and cohesion of the community, in the service of the innovation process.

However, there are at least two questions which have no response, probable challenges for the future of the "Créativ'Café" or more generally the ideation process:

- Is it necessary or desirable to mark the process by "campaigns", as with the innovation process? This organization could lead to a transformation of the process opened in *stage-gate* and therefore an assimilation into the already existing innovation process.
- The opening of the ideation process towards, on the one hand, all the occupations of the firm (internal extension) and, on the other hand, the partners of the Schmidt Groupe (external extension), whether customers, suppliers, sales networks, etc.

The responses to these challenges should *in fact* be consistent with a more general corporate strategy concerning opening, internationalization and, internally, freedom of initiative for employees.

The Créativ'Café: Afterthoughts 2 years later*

Upon its launch in 2016, the "Créativ'Café" was well received by the company's creative community, in response they showed strong enthusiasm which fuelled the initiative with great energy. As could be foreseen, after the initial wave of hype, interest for the initiative started to wither and it became necessary to devise new mechanics to keep it efficient in the long run. One would naturally think the best way to maintain interest for the "Café" would be to communicate on its successes, however, developing new products is a long process, leaving a gap to fill between the hype and the first glorious achievements. We call this gap the afterhype, we will address how to manage it in the first part of this addendum.

Then comes the question of what to do with projects once they got approved by the decision-making committee, namely: when can they integrate the company's classical stage-gate product development process? Too early and they will not be defined enough, overloading engineering and marketing departments, too late and they will have been forgotten or surpassed, wasting the time and resources invested in them. This delicate phase is the landing of the project and will be discussed in the second part.

I. Managing the Afterhype

The "Créativ'Café" is an incubator aimed at developing collaborators' ideas for new products. When it was first launched, ideas abounded as some colleagues had been nurturing their project for a long time, months, even years, others came up spontaneously with ideas in order to participate actively to the Café. Quite naturally this flow of ideas ran dry after a year or so and we had to devise ways to keep the initiative alive. As a first attempt, we defined a number of fields of innovation to investigate, in compliance with the long-term strategic goals of the company, and periodically issued calls for ideas in those fields. Said fields could be very narrow, e.g. "storing solutions for tiny living spaces" or quite wide, e.g. "the digital kitchen". We hoped those suggestions would entice participants with a particular interest for the subject (e.g. a person spending holidays in an RV or one with a strong taste for domotics ...) to come up with new ideas. Conversely, highlighting some fields meant excluding others and if this were to be perceived as a restriction of creativity, it could stop some members in their tracks.

**Tristan Cenier*

Those attempts were unsuccessful, no idea emerged from them. Through discussions with members of the creative community, the facilitator discovered that interest for "Créativ'Café" had already waned enough that collaborators, although still interested to participate, would not go as far as coming up with ideas. The question of whether suggesting fields of innovation was of help or a hindrance remained unanswered.

As a second attempt, we reasoned that instead of calling for ideas, we could rekindle interest in the "Créativ'Café" initiative by setting up a pipeline of ideas to inject in the incubator. Sources for the pipeline were sociological watch, collaborations with entities outside the company: architects, schools of design, local associations, but also in one instance an idea that emerged from a past "Café". Setting up this pipeline was an easy task, the difficulty lied with the company's creative community. How would they react to ideas that did not emerge from within? We discussed earlier the importance of the sponsor for a successful collective-intelligence process. The sponsor is both a centre of mass with high gravity, giving the idea a face, an identity, coalescing collaborators around it, and a drive that makes sure the creative process is carried to its end (or at least up to the decision-making committee). Injecting an idea from the pipeline raises two complementary problems. First problem: it means that the idea has no sponsor, henceforth building little — if any — traction. Second problem: since he is the one injecting the idea, the facilitator will *de facto*, albeit falsely so, be perceived as the sponsor and we know that they have to be two separate roles, in the same way one cannot be judge and jury.

To summarize, after the initial pool of ideas ran dry we resorted to injecting ideas from a pipeline, thus creating new potential pitfalls: lack of motivation due to the idea having no face, so to speak, and confusion arising from the facilitator being perceived as the idea's sponsor. Reality proved both pitfalls right and revealed a third one. To overcome the lack of sponsor, the facilitator worked extensively on the ideas to make them look "sexy" and sell them more easily to the community. This work included 3D models of potential applications for the ideas, along with description of goals and hopes for each. Ideas were then proposed to the community with a call for participation. Three ideas gathered enough interest to start the incubation process but none reached a satisfying end. Participants would show up, be focused for ten minutes until they started to lose focus and share personal anecdotes, in relation with the subject but totally sterile. "Créativ'Café" had become more of a social club than an innovation machine. In each instance, no exploitable result was obtained after the initial 5 or 6 sessions. At that point, the facilitator stopped scheduling more sessions to test attendees' motivation, they never asked for more, a clear indicator that the will to go on was lost. To be noted also was the poor attendance and frequent tardiness, more indicators for a lack of drive in the project. This was somewhat surprising since none of the attendees were slackers, some of them even were very active participants in previous "Cafés". Later, attendees provided very interesting feedback to the facilitator. The failure of the setup could be attributed to several factors that all sum up to one significant deathtrap: a lack of challenge. First and main factor: attendees felt that most of the work had already been done upstream: the idea was already there and well documented, there was little left to do, or only the less creative part (i.e. the business model). Second factor, a biased role of the facilitator: whenever the latter encouraged attendees to challenge or criticize the work in progress they disengaged from the process, whereas they never hesitated to challenge the idea of a sponsor. A plausible explanation for this is that when an idea is injected, challenging it means challenging the facilitator, leading to potential conflicts. The lack of challenge stemming from those two factors translated into a lack of motivation and there was no sponsor to pour energy back in the mix. This shows that ideas should be kept as raw as possible before submitting them; the facilitator should adopt a lazy stance upstream of the "Café". More importantly, it highlights once again the importance of the sponsor.

For the third attempt, we thought "instead of injecting ideas into the community, manage the community to make it come up with its own ideas". To that end, we devised a two-wave initiative. First wave: pick a subject from our fields of innovation and organize sessions to simply discuss it. The discussion would remain quite shallow, focusing only on very concrete aspects of the subject and hopefully yield a batch of fresh new ideas that will in turn spark multiple "Créativ'Café" in the second wave. Most importantly, those ideas will have been produced from within the creative community and it is a safe bet that attendees of the first wave will constitute themselves sponsors to carry the ideas through to the second wave. Colloquially, one could say that the first wave is incubation of sponsors while the second is the more classical incubation of ideas. The experiment is currently ongoing.

II. The Landing

Schmidt Groupe develops new products in short cycles, following a detailed and somewhat rigid stage-gate process, divided into three main stages: ideation, pre-project, project, the latter meaning industrialization and commercialization. Gates are crossed under supervision of the company's executives and detailed documentation is recorded as it moves forward. When managers arbitrate how much human resources they'll allocate to a given project, they refer to that documentation to make their decision. In consequence, a project entering the pipeline will eventually come out and reach the stores (or be delayed if workload capacity is reached), whereas a project not in the pipeline will get zero resources. The result is a very effective process, but at the expense of agility.

As we saw before, the "Créativ'Café" is a very early stage ideation process, aiming at restoring agility and unhindered creativity in the innovation process, as well as long-term thinking. Projects that receive the decision-making committee's approval will eventually need to enter the official pipeline. The question is when. Here again lies a delicate balance: enter too soon and all agility is forfeited, enter too late and the project will suffer from a lack of resources.

1. *1, 10, 100, 1000*

Innovation is by nature risky business. Developing a product demands financial resources and the more uncertain the project, the higher the risk those resources will go to waste (project never reaching the end or product turning out to be a failure). Anyone who has ever founded, or worked for, a startup knows that no matter how promising your project is, you will never find investors willing to fund it from beginning to end in a single installment. To mitigate risks and potential losses, the guide rule is to invest in small incremental steps, carefully assessing progress in between increments. This process can be summarized as such: "I, business angel, will give you 1€ now, let's meet again next week and you show me what you have done with my euro. If all goes well, I might give you 10€ to go on for another week, then a hundred and so forth".

Most companies while creating new products stick to their core skills and field of expertise. Most of the development is therefore mastered and it reduces risks considerably. Not investing in risky projects however means your catalogue of products will stagnate and be outclassed eventually by competitors. Here again, adopting a 1, 10, 100, 1000 approach will mitigate risks and allow the company to innovate, with R&D in the role of the startup and the CEO investing some of the company's internal resources, keeping the investment small at first to reduce loss in case of failure, then investing more as uncertainty decreases.

"Créativ'Café" is clearly the 1€ step: very little resources were used to design innovative new products, the initiative is fuelled mostly on good will and personal efforts. After validation, it was time for the 10€ step: sprinting.

2. *Sprint*

Sprinting is a classic in creativity management: concentrate effort and money in a project over a short period. The sprint team will differ from each project and is composed of the facilitator and attendees to corresponding "Café", on a voluntary basis. Said attendees accepted to devote one full day every fortnight over a period of 5 months (10 sessions), with the goal to develop a fully functional version of the product designed in the first step. Sprinters where willful and well disposed, ready to give some of their time, even if it meant working harder or longer other days of the week to catch up on their regular workload. Unforeseen difficulties (availability of tools or materials, necessity to share the workshop, elements broken in between sprint days…) however slowly eroded their spirit until they no longer felt it worth the extra effort. With this type of initiative, a tacit contract is established between participants and the facilitator: "we give you some of our time and you make the best out of it". On numerous occasions, participants felt the contract was not fulfilled. In consequence, sprints where turned to shorter cycles: one full day every three weeks over three months (5 sessions) goals are to be carefully stated for each cycle and necessary resources to reach that goal must be assessed. At the end of a cycle, progress must be recognized by top management, in a committee or any other administrative device, as well as the participants' efforts. If progress is satisfying, the next cycle can be approved in that committee, as well as allocation of resources.

Repeated approval from top management and resource allocation are necessary elements to avoid the worst pitfall for projects developed outside the official project pipeline: the hitch-hiker effect.

3. *The hitch-hiker effect*

When driving your car, you are likely to pick-up hitch-hikers along the way. As long as you can just drop them off somewhere along your scheduled route the experience may prove pleasant as you can chat along the way and spend an overall better time than sitting alone in your vehicle. However, when the hitch-hiker asks you to make a detour, has too much luggage, puts mud on the dashboard and eats your lunch, however pleasant the conversation, you will probably not pick it up the next time.

Any projects that are not in the stage-gate process, as is the case of projects issued from "Créativ'Café", are hitch-hikers. Individuals and managers will probably agree to allocate a few resources or spend a little time here and there to help develop said projects, even more so if they were participants in the "Créativ'Café", as long as it is not too demanding. Whenever a threshold of time or effort is reached, they will refuse to help. Who can blame them? They have goals to achieve, deadlines to meet, eight hours in a work day and a personal life the rest of the time.

Interestingly, as soon as a compensation system is set up to reward the driver (i.e. sharing costs for gas, toll, mileage, etc.), hitch-hikers are no longer perceived as such, rather they become fellow travellers. As proof, carpooling initiatives are blooming, and the French company Bla-bla-car is a strong example.

Within the company, the compensation system can take the form of creative slack, a portion of their work time that employees can — and often must — use to work on side projects, as is the case with Google or 3M. A simpler form is simply that each manager will have received word from his own manager that eventually she/he will have to allocate resources to a project even though it is not in the main stage-gate pipeline.

III. Integrate

Innovative "on the side" projects are kept alive thanks to unusual resources in the company. A transient taskforce emerges from the underground and carries the project through thanks to good will, enthusiasm and additional effort. It demands agility, audacity and creativity. To preserve this, a number of initiatives — the middleground — must be set up to accompany the movement and make sure momentum builds up and is not lost. The main pipeline, too rigid and too risk-averse, is not fit for that kind of project, at least not until a late stage when perceived risk has been reduced to an acceptable level. But with no supporting frame, day-to-day managerial preoccupations will act as a handbrake on the projects. It is therefore necessary to devise cycles of investment, involving the upperground, which will serve two purposes: give the projects a seal of approval from top management, signalling "it is okay to support this particular project by any means you can spare" and making sure necessary resources are available in order not to waste participants' time and energy.

As a matter of fact, this is another stage-gate process, in this instance designed specifically to address the risky nature of innovative projects and to the management of a creative community. Eventually two pipelines will coexist, one traditional, close to the core and effective, the other one more innovative and agile, able to diffuse a future-oriented, set-for-innovation-mindset in the company.

References

Cohendet P. and Simon L. (2015). Introduction to the special issue on creativity in innovation. *Technology Innovation Management Review*, *5*(7), 5–13.

Lerch, C. and et Burger-Helmchen, T. (2015). La créativité décentralisée: le cas d'une entreprise industrielle personnalisant ses produits. *Innovations* n°48, 3, 107–127.

Part 3

Communities of Innovation in Industrial Ecosystems of Multiple Organizations

© 2021 World Scientific Publishing Company
https://doi.org/10.1142/9789811234286_0007

Chapter 7

The Renault Innovation Community: 2007–2017

Frédéric Touvard and Dominique Levent

At the beginning of the 2000s, the Internet is developing, digital invades our lives and colonizes all our technical systems. For the automotive world, engine manufacturer and sheet metal bender in its DNA, it is a huge wave that is already overwhelming us. A wave that comes in addition to two others just as strong:

— on the one hand, climate change and the increasing scarcity of raw materials;
— on the other hand, the emotional value of the automotive object which oscillates between pleasure, status and convenience inseparable from our modern lives.

To ride these waves, we would have to change, and quickly; we would have to understand the digital world, its potential, its culture, its management modes, but also to get out of our comfort zone and beyond our borders. We had to go out into the world that is changing, open up to new partnerships and accept to shake up our visions and beliefs.

In 2007, what better place than California to plunge into the heart of the digital world! A simple excursion quickly convinced us of two essential transformations to be made:

— learn from our mistakes, not considering them as failures;
— share our visions, our ideas, and accept to push them, to be enriched by the ideas and skills of others.

It is in this context that the Renault Innovation Community was born, created by Yves Dubreil, former Program Director for Twingo, and Dominique Levent, former Product Manager for the first Scénic and Kangoo.

The secret equation of this Community is to "waste time" intelligently with the various "accomplices" of the colorful world of innovation:

— to recruit talented people who are comfortable manipulating doubt and not companies that are a priori useful to our business;
— collect authentic testimonies, share them and question their meaning through contradictory but benevolent debates and put them into perspective through the eyes of a philosopher who is aware of the recurring cogs and wheels.

To live this adventure, Yves Dubreil and Dominique Levent have recruited Frederic Touvard, a coach and specialist in "issue-based project management", and Dominique Christian, an "on-call" philosopher (storytelling and strategy) and a Chinese landscape painter.

And it is with passion that they have animated this Community for twelve years with around thirty members at its launch, and more than one hundred and fifty in the last few years. One third of the members were Renault innovation players, another third were innovation managers from other very diverse companies (Valeo, L'Oréal, SNCF, Air Liquide, Safran, etc.) and a final third were academics and consultants (Ecole des Mines, Ecole Polytechnique, Strate College, etc.).

This Community has met three times a year in plenary sessions to discuss strategic and potentially transformative topics for our practices and businesses during more than 30 sessions; topics such as the collaborative economy, the relationship between large groups and start-ups, the autonomous vehicle. These plenaries were initiated by a first session in a small committee to collectively bring out, with the help of philosophy, the important questions or contradictions to look at in order to better understand the stakes of the theme. During the plenary session, testimonies, sometimes contradictory, enriched the points raised and sometimes even a "project boxing" session closed the debates to offer the Community and the project leader a real anchorage for our reflections.

…..

This beautiful adventure ended at the end of 2019 for budgetary reasons (Renault's very difficult economic context), but perhaps not only that. Indeed, the time had probably come to move on to something else because the context had changed a lot since then. All the players in innovation have mobilized to help their companies get through crisis after crisis, transformation after transformation.

The time for mutual learning is more necessary than ever. If Renault has carried this vision to that point, it is now time to adopt a more active and rapid approach with extended communities, perhaps fractal with rotating piloting …

II. The Origin of the Emergence of the Community … the Old Regime

1. *Past initiatives on project management*

Understanding Renault's innovation community means looking at it through the prism of past initiatives — failures or successes — that led to the emergence of this structure in 2007. However, the first steps of this project were made in the late 1980s with the Montreal club created by a small core group of people from Renault and outside companies. In 1989, Raymond Lévy (Chairman and CEO of Renault from 1987 to 1992) set up project departments to avoid waste and make the silos between the different departments more fluid. Yves Dubreil was given full powers when he became Project Director on the Twingo, but as a novice, he did not know how to achieve this objective. So, he worked with Gérard Dubrulle — formerly HR Director at Renault — to find answers to his questions. Dubrulle had already begun to reflect on project management and the link with innovation by gathering information on the subject. He wanted to bring together practitioners and theorists in order to set up training courses to make managers aware of this new form of management necessary for this new organization. The founding members of the Montreal club[1] — including people from Renault and outside companies — met for the first time in Montreal, Canada, to observe how North American companies operate. Each member of the club had a value equal to that of his or her neighbour that allowed them to express themselves freely, thus creating "an Utopia of self-organization or even the absence of organization".

In addition to being a space for training and debate, the Montreal club was also a place of communication that made it possible to present projects to an audience of experts. A real communication plan based on the expertise of club members was put in place for the launch of the Twingo and contributed to its success: it in turn strengthened the club's reputation. While the retirement of the most active members and the difficulty in finding a balance between innovative and standard projects led to its closure, the Montreal club remains the leading innovative community in terms of both its management method and its purpose, inspiring other community and management models.[2] Yves Dubreil sought to renew this original model in the mid-2000s. During an afternoon session, he invited some 50 people from Renault and members of outside companies to share their original project experiences. Philosopher Dominique Christian was also present to comment on these experiences and open new windows of thought. It was a question of "losing an afternoon's work but in an intelligent way" to better resume one's activity after the exchange. This event was also a success. The innovation community as we know it today is a clever balance between these two events that are the Montreal club and the afternoon of exchanges and comes in response to a problem of the 2000s: saturation. Saturation reflects the feeling that it is impossible to evolve, to change the way we create, produce and think about creation at the same time.

---

[1] See https://www.cairn.info/revue-le-journal-de-l-ecole-de-paris-du-management-2004-2-page-15.htm.

[2] The Montreal club has experienced a split between a group supporting standard projects and another focused more on innovative projects (although club members refer to them as "one shot" projects). Awareness of the need to know how to do innovative and standard projects has occurred, but late. Indeed, standard engineering allows a lot of savings to be made in order to be able to finance innovative products. It is this loss of vision and shared values that led to the closure of the club.

2. *The context and the triggers: The learning expedition in the USA*

It is a scourge that necessarily affects all businesses: "the monster of habit". It freezes people in practices and postures that hinder creativity. Indeed, companies have nowadays internalized hyper-rationalized production methods leaving little or no room for innovation. This rationalization appeared to be a necessity, a guarantee of quality, a value on which the manufacture of Renault models is based. Today, however, quality — although essential to meet customer needs — is not enough to ensure their loyalty in the face of an ever-expanding offer. Innovation is seen as a means of coping with an ever-changing world. But, paradoxically, it is precisely because we reach the end of something that innovation can be born. It is a response to an emerging problem, to a complicated or even inextricable situation that requires its contribution. In this sense, it becomes at once a solution, a means and a goal for development. Innovation is a process of regeneration. It allows companies to become one with the need to evolve imposed on us by our society. If we remain trapped in a habit that reflects a refusal or fear of change, it will lead us straight to another scourge: obsolescence. This reality is easily illustrated through the examples of Betamax, Kodak and Virgin, which are a sign that quality is never enough if it cannot be combined with creativity. Sometimes unable to follow or accept change, many recognized companies are forced to close their doors.

This tension between extreme rationalization and the need to generate creative space is the starting point for a broader reflection on innovation initiated by Dominique Levent, Director of Innovation at Renault. Like many other groups, Renault is not exempt from the need to "do things differently", a necessity already imposed on the other side of the Atlantic. At the request of Yves Dubreil, who at the time was seeking to facilitate the opening up of Renault, Dominique Levent went on a "learning expedition" to California in 2007 and discovered another way of creating that was more collaborative, more efficient and more focused on exchanges. Before its alliance with Nissan in 1999, Renault, like many other structures, was too self-centered, which could have been detrimental to its ability to compete internationally. In San Francisco, several points caught Dominique Levent's attention:

— the valorization of the failure which appears as a pledge of maturity of a reflection and/or a project;

— the porosity of professional barriers: all the services communicate between themselves and with the outside world;
— the informality of decision-making, which can be done during a barbecue or an evening out.

On his return to France, Dominique Levent was looking for a new way to promote innovation. In 2007, he met Frédéric Touvard during a lecture at the Paris School of Management, who shared the same intuition. We can really talk about intuition because both of them felt the need to regenerate practices without knowing exactly what or how. At that time, Frédéric Touvard was Director of Operations for an internal Air Liquide spin-off called Axane, in charge of developing the hydrogen fuel cell. The way this spin-off is managed and creates innovation makes it a "Gallic village" in the midst of the structured giant that is Air Liquide. It has thus become both an object of disruption and covetousness, in the sense that by refusing to follow highly codified rules, it has imposed itself as an actor of change. Frédéric Touvard will find this dual status in Renault's innovation community. Yves Dubreil and Dominique Levent then suggested setting up an office in California or an innovation community in France. In the context of the 2008 crisis, the office was not selected, but the innovation community aroused interest. With a shared vision, Dominique Levent and Frédéric Touvard, with the help of the philosopher Dominique Christian, would build what would become the Renault Innovation Community. It is this meeting between the technical skills of the engineer and the philosopher's thinking that would be the foundation of an innovative approach to innovation, which today more than ever appears to be a real necessity.

The innovation community from its inception crystallizes this need in the sense that it was created not to give clear and structured answers, but to answer constant questions about innovation, or more precisely, what makes sense in innovation. All the more so as innovation takes on several forms and subjective definitions. Being innovative in innovation means first of all not taking such a hybrid term for granted, but also knowing how to bring it to life through words and deeds so that it can play its full role, which is what Renault's innovation community is doing. Although the foundations were laid very early on, it is the departure of Frédéric Touvard from Air Liquide in 2010 that would allow this community to grow. Indeed, he then became the community's leader and development manager, supporting Dominique Levent in his role as director. The principle of this community is to create another collective way of innovating. Innovation is defined as "creating something new that generates potential value". It can be social, economic, spiritual, intellectual, personal and collective at the same time. The community is a catalyst for all these values, which it questions. It is in this sense that it is truly a pioneer in France.

III. The Birth of the Community

1. *A first circle*

The innovation community has been built from the outset around the idea of decompartmentalizing existing uses and models in order to benefit from a richer and broader contribution. It is therefore natural that it has relied on members of the Renault group, but also on individuals from other companies or recognized academic institutes, to which a philosophical breeding ground has been added. This first circle was made up of people who were fully convinced of the importance of creating an ecosystem that would encourage meetings dedicated to innovation issues and who, for some of them, had already participated in previous innovative projects such as the Montreal club. The confidence of the members towards this community and the other participants was and still is one of the prerequisites for the smooth functioning of this community. It is this trust that will allow the emphasis to be placed more on sharing and listening than on the need for production. Through this rule of trust, "everyone benefits" because it allows to maintain the network, to meet people with the same interests, to benefit from an "oxygenating" content, to identify new ideas and to strengthen the links between the different stakeholders.

This connection between these different networks allowed the emergence of a first plenary session which was based on a small circle of 25 to 30 people drawing on both scientific and literary resources and which highlighted the importance of this dual methodology. For, and while common opinion tends to see them as two worlds that do not communicate with each other, it is precisely this encounter between the technical skills of the engineer and philosophical thought that would be the breeding ground for an innovative approach to innovation that already appeared to be a real necessity, and which is now more than ever. To create the conditions for this, the increase in the number of members and exchanges are important, because it is they, with their different experiences and profiles, who will allow creative friction. Thus, in 2009, the community welcomed Californians met by Dominique Levent during her learning expedition. International exchanges are also one of the ways to get to know emerging concepts that make sense abroad and that could be used in France. It is also an opportunity to exchange on the respective advances of the innovation groups and to unblock complex situations by drawing on other experiences.

In order to consolidate and prevent the physical and intellectual dispersion of this nascent community, it was necessary to refocus resources, knowledge and efforts around two terms — objects that make all the more sense as they correspond to issues that Renault and the other member-companies are confronted with on a daily basis: innovation and mobility.

2. *The objects of the community: Innovation and mobility*

Belonging to a so-called innovation community means first of all questioning this overused term and taking into consideration — with a view to enrichment — the vision of each member. Thus, they do not hesitate to challenge each other to come up with more fertile ideas from these debates.

Mobility obviously made sense with the identity of the Renault structure. But it is not just about cars, because here it is understood in the broadest sense, defined as everything that has to do with movement, traffic and mobile life. It is therefore natural that this open and dense community is also interested in the contribution of other groups concerning these everyday themes such as Google, Uber or Amazon. Each in different ways and in different areas responds to the challenges posed by our society in terms of mobility and circulation.

Thinking about mobility also means doing so through the prism of immobility, which is still little studied. To be mobile is not only to move, it is above all to move in a space and to interact or not to interact with objects and people. The opposition between the individual vehicle and public transport underlines the importance of grasping the whole: space vs. isolation, opening vs. closing, collective vs. individual. Indeed, owning one's own vehicle includes a certain comfort in having a personal living space, while being aware that the car itself does not have its own space to move around. On the other hand, public transport, which sometimes allows for greater mobility, is the scene of a desperate search for that personal living space. Indeed, one only has to look at people glued to their mobile phone or book to understand that transport has only their name in common. Innovating in mobility therefore also means anticipating human expectations and behaviour and the evolution of spaces. It also means realizing that more and more people are opting for an "in-between" solution, which is vehicle-sharing. From these new concepts of mobility emerge new questions: how to find pleasure in not owning a car but only using it? How to succeed in finding one's living space? The automotive sector, and more broadly mobility, is being pushed to evolve so as not to die and not to reproduce the production patterns of the 1950s–60s, because the future competitors or principals of the automotive sector will certainly not be car manufacturers but more broadly players in mobility.

Thus, these two terms — mobility and innovation — are "fashionable" objects that appear to be both the problem and the solution. Problem, because it is true that they challenge us on a daily basis by forcing us to rethink our lifestyles according to other ecological, social, economic issues… Solution, because they also remain the answer to these problems. It is therefore especially important to cross-reference these two themes in order to obtain more solid and relevant content. Mobility without innovation is doomed from the outset to failure. Innovation without innovation is just an empty word.

3. *Rituals: Density, place, language, tone*

A community in the sense of a set of individuals bound together by common expectations and values can only be welded together by a series of elements that, when put together, constitute what can be called rituals. Rituals create a sense of security reinforced by the idea of contributing to the birth and maintenance of traditions. Rituals aim to:

— to create a sense of belonging and recognition, in that they allow each member to become one with the rest of the group;
— strengthen the identity of the community by setting a clear framework;
— ensuring its proper functioning.

For example, plenary sessions are an integral part of these rituals because the place and manner in which they are organized helps to maintain this sense of identity. Held three times a year at the Ecole des Mines de Paris, they fully assert this Parisian identity. These events bring together 80 people exchanging only in French, to be able to debate in "true speaking", without ambiguity. What they have in common is their desire to question what makes sense today in a changing world. These meetings follow a well-defined process. Indeed, sequences of speakers are organized to provoke debate and reactions. The presence of two philosophers, Dominique Christian and Thierry Ménissier, reinforces this impression of a framework without it being completely fixed. Indeed, the two philosophers will come to challenge or enrich what is said by bringing, over the course of the sessions, offbeat elements. For example, while Thierry Ménissier introduces the subject in a dense and structured way (a lot of information), Dominique Christian will give in conclusion elements of stalling and breaking. These sequences end with practical workshops during which the participants take ownership of the elements brought during the interventions. In this way, idle time is avoided by ensuring a dense rhythm that creates the conditions for richer reflection. It is precisely this density that current and future members of the community are looking for because they know that they will have access to varied content (readings, quality speakers in different fields, tools…), inserted in an environment of perpetual questioning that will allow them to think differently.

After a stint at Renault, Anne Bion, now Director of Innovation at Nutriset,[3] looks back on her experience in plenary sessions. Her team was in charge of steering a research programme on new business models for Renault and, in doing so, she took part in several plenary sessions, notably to find avenues for reflection on the issue of "service mobility". For her, they are a place for extending and enriching the initial questions that can be asked outside the company in both professional and private environments. In this sense, the plenary sessions contribute to an "exploratory dynamic" that is necessary when working on innovation. This exploratory dynamic questions participants as individuals and as potential innovators. Its objective is to allow all members of the community to reappropriate on a daily basis the methods put in place during these workshops in order to answer these questions: how to create differently, why to create differently.

[3] http://www.nutriset.fr/fr/accueil.html.

IV. The Development of the Community

1. *The rules of engagement and functioning of the community: Benevolence and exigency*

The number of members present in the community is an essential element in fueling this need for density because it is their meeting that creates the critical mass, the foundation of collective emulation. This community is collective in essence, bringing together different profiles from a wide range of Renault departments, including Human Resources, Research & Development, Purchasing and the Innovation Department, as well as from outside the company. Today, more than 150 people representing nearly 40 companies* make up the community. In addition, there is a strong connection with the academic world, particularly through the presence of the Grandes Ecoles, as well as philosophers, historians, sociologists, literary experts and PhD students. Other so-called "hacker" events underline this broad openness. They can attract the attention of the community. It is thus grafted on to external actions allowing the provision of innovative, more gourmet food based on a common synergy.

Recently, a conference on disruptive innovation in France was organized by the community on the occasion of an event set up by Shamengo.[4] This multifaceted organization is a breeding ground for the exchange of concepts and ideas as well as members' projects. Each member contributes to the emulation and friction necessary for the creative emergence around a double relationship of benevolence and requirement. This duality authorizes freedom of speech, which gives the right to controversy, in other words, the possibility to talk about things that disturb or may be in contradiction with general opinion.

- Each individual who joins the community must be able to count on the fact that the other members will only judge or criticize his ideas, never his person. But these criticisms will only have the sole purpose of bringing out more relevant issues and thus benefit the entire community.
- This can only be done on the condition that you play the game and bring elements to enrich the exchanges. Each member is both an actor and a spectator, he or she cannot be one or the other. Integrating the

[4]See: http://www.shamengo.com/.

community is therefore about finding a balance between giving and receiving, between benevolence and demands, rules that commit the whole community, where each member must be aware of the key role he or she plays for others and for the community, and that consequently he or she has as much to learn from others as they have to learn from him or her.

— Thus, it is impossible to join the community with a profile of "lesson giver" or "Mr. Know-it-all" because not only would the individual not be able to fit in, but the whole community would turn against him or her in a movement of rejection. We share, we don't impose!
— In order to contribute to this necessary exchange and to an openness that allows each member to feel confident, no legal contract concerning confidentiality or intellectual property binds the members of the innovation community. Each member comes in this sense with his own responsibility concerning what he is going to exchange. Based on mutual respect of the word, these exchanges, ignoring any cumbersome legal system, promote "true talking" and the sharing of visions and doubts.
— Collective creativity sessions can contribute to the emergence of ideas with high commercialization potential, but these, when identified, are conducted outside the community. In this way, project teams can be created outside the community. Their maturation can only take place outside so as not to "pollute" the functioning of the community, which must remain a space for free reflection rather than for designing operational projects.

These rules of engagement are the rules on which the recruitment process is based. More than a typical profile, it is an identity that makes sense with the rest of the community that is sought first and foremost. Recruitment is therefore intuitive, through discussions and a test invitation to a plenary session. It is a question for the newcomer to understand how this atypical community works. This recruitment model allows the person to be hired more than the company. So even if a person changes companies, he or she remains a member of the community if he or she wishes. The community's "vocation is to create a rather hybrid space to encourage slightly different ways of thinking at Renault so that people at Renault who already have this spirit can find an echo. The idea is to bring together individuals with different profiles but who share common values to create a genuine culture of innovation. There is only one selection criterion, but it is essential: "to correspond to the culture of this community", in other words, to share the posture and to want to learn and exchange around new methods, attitudes and innovative tools in a logic of questioning and doubt.

2. *Doubt and situational awareness as conditions for emergence*

The innovation community has been based on doubt from the outset. It must doubt in order to exist. And it is precisely the trust placed in the community and in the individuals that make it up that allows doubt as a creative premise. Indeed, it allows one to question one's own questions and visions of the world but also the community itself. Doubt participates in deconstruction, as a weapon to combat false evidence and never to take for granted what is. As Dominique Christian explains, it takes two incompatible, opposing reading grids to start "thinking" about complex subjects. Doubt in the sense of a questioning drive is therefore a key tool in Renault's innovation community because it is the starting point for richer reflection "off the beaten track". Thinking "off the beaten track" means deconstructing facts in terms of innovation. It can be the creative process or the posture that the innovator can adopt. In either case, the goal is to find better approaches through richer questioning that will not necessarily lead to absolute certainties. For example, here are some questions that were asked during the plenary sessions:

— What in an experience is unique to the person experiencing it?
— In what way is an experience, even an ordinary one, necessarily transforming?
— From an ethical point of view, what part of an experience is recommendable, what part is not, and according to what criteria?

Plenary sessions are always prepared with little advance preparation, which makes it possible to draw on recent developments in innovation and mobility. It is important to leave space to grasp new concepts as the plenary approaches. This freedom explains why it is possible to show "situational intelligence", i.e. to adapt to the needs of the participants and to change the format or duration of a workshop or intervention during the event when it is perceived that the attention and energy of the participants is focused on particular topics.

3. *Prototype projects and themes*

During the development of this community, a first phase was based on the use of numerous prototypes. For example, a first card game on scenarios of mobile life use. The cards represented people with different profiles, means of mobility, obstacles and proposals to solve problems. The arrangement of these maps made it possible to test various scenarios in workshops. It was a playful and concrete way of confronting the challenges posed by mobility.

In 2010, a scenario on electric vehicles was also set up. Jean-Marie Réveillé from Renault lent 5–6 electric vehicles which were used during a plenary session during which several scenarios were built. There were two teams per vehicle that joined forces and the scenarios collided to create new situations. For example, people had to get into a self-piloted electric vehicle, but one was ill and the other had to go to the doctor. What happens when an unregulated mobility object has to be shared? The aim is therefore to bring out typologies of situations that could generate innovative or problematic elements. From these scenarios, a synthesis was created which, in half a day, highlighted a few situations of use of electric vehicles.

Other so-called experimentation workshops were organized to test techniques and creativity tools such as serious play legos, C-K theory, Bono's 6 hats…

Finally, Dominique Christian has worked hard to try to make philosophy more accessible. He created what he called "a philosophical furniture", which is a series of posters on the theme of mobility, but through filters or authors: Montaigne and mobility, imagination and mobility… are examples. This led him to write a book called *La mythologie de la mobilité* (*The Mythology of Mobility*). These posters were also used during a seminar at the Cité des Sciences at the beginning of 2012 for an exploration workshop on the subject of mobility. They were also the subject of a travelling exhibition in companies in the community. It is important in the development of the community to capitalize on the productions of its members. Each member of the community who has published a book that makes sense with the objects of mobility and/or innovation can present it during plenary sessions and Dominique Levent often offers a copy to the participants. This allows for an exchange around the productions of the community itself in the community.

V. Maturity

1. *Projects, events and networking*

The innovation community is a structure that allows "to impose itself as a complement to the internal initiatives deployed by companies in order to feed them with offbeat content and original practices". The innovation community is therefore a means of supporting the wider community of Renault and even all corporate communities. It is a means of development in the sense that it enables old practices to be linked to new thinking. The aim is therefore not so much to provide perfectly structured answers as to be surprised by what is, and to reflect on what should not, but could be. The community thus appears as a "space of interpretation" highlighting what are called "weak signals". These can be defined either as new ideas, theories that are new but not necessarily talked about, or as fashionable concepts that need to be revisited. In either case, what will be important is to understand the mechanisms that have led to these concepts in order to better dissect them and grasp their limits or, on the contrary, their potential. For example, one debate that the innovation community has focused on is the connection between large groups and start-ups, with which more and more large groups are forging links. One of the main aims of the debate has been to understand the impact on innovation that this type of relationship can have. The innovation community deals with current economic, political and social issues. This is why the plenary sessions dealt with rich and varied themes such as "The collaborative economy", "Vices and virtues in innovation", "The breakdown of desire", "The loneliness of managers", "Cities of the future", "The autonomous vehicle" or "Open source". We thus find a balance between more theoretical and more practical subjects, which reinforces the quality of the questions by anchoring them to reality.

The innovation community also lives through other events and projects that can be spread over several months. For example, the PL-U-I-E project (*plateforme urbaine intermodale d'échange en énergie électrique,* i.e. urban intermodal platform for electric energy exchange) is one of the federating projects that was set up at the end of 2011 and has reached maturity. It consists in modelling with the real quantities of energy present at an intermodal station with subways, electric vehicles, pedestrians walking on piezoelectric slabs, buses, bicycles, the production of renewable energy (solar and wind). The idea was to understand the energy consumption of a station over a day and a whole year and which deposits could be released. For example, the electric vehicles that would all be coupled to the station at night could be used to feed the power grid to get through the peaks in the morning.

2. *Sensitive devices: The Distillery and the "boxing"*

— *Philosophy and innovation: The Distillery*

In parallel to the plenary sessions, another concept would emerge in 2013 that was born from the desire to use philosophy as a resource. It is rapidly structured around 5 to 6 targeted themes per year in relation to innovation and management. In a half-day format, participants address these themes through philosophical workshops following the intervention of thinkers Dominique Christian and Thierry Ménissier, the latter having joined the innovation community at the time of the creation of this new concept. The words of the philosophers are brought in a structured and "top-down" way between knowing and learning. This organization very quickly shows its limits in the sense that there is too great a gap between philosophical concepts and their operationalization in companies. This dichotomy leads to an uncomfortable relationship between the thinkers and the people present, who from one workshop to another are often the same.

The aim then becomes to reorient the group dynamics by making this new concept more concrete and operational. Philosophy maintains its central place because it allows a step back from the experience of humanity, which is necessary for a richer reflection and brings a new look and more relevant questions. However, in order for the philosophy to be used by business professionals, it is put in relation with the plenary sessions. Thus, still with the aim of making the philosophical contribution more operational, it is decided to set up exercises of about thirty minutes which will be entry points for approaching the themes in a more playful and direct way. For example, during the workshop on the subject of "vices and virtues in innovation", the participants in pairs, after having found a vice and a virtue, had to push the virtue to the extreme in order to turn it into a vice, thus demonstrating that any quality in innovation is relative, depending on the people and the situations in which they find themselves. Following these exercises, the philosophers develop concepts by mobilizing authors and references that will be as much material for more dense reflection during the workshops.

These workshops reinforce the operational character of the Distillery because they allow each participant to produce concepts and to rework creative paths initiated inside or outside these workshops in the light of philosophical contributions. For Thierry Ménissier, the interest of the Distillery lies in the relationship of equality that is established between the members of the community and the philosophers. Far from adopting an overhanging academic stance, the latter remain vigilant in creating the conditions for a liberation of the philosophical word that allows each person to feel capable of taking a position on elements of deeper meaning. It is in this sense that we can truly speak of workshops and not conferences. The Distillery is a unique and practical concept that cannot be found elsewhere.

Today, one wonders whether Renault's innovation community is not tending to become the Distillery. While the foundations of this concept are well established and solid, its name is only very late in coming. It was during a working meeting between Dominique Christian and Frédéric Touvard, who discussed the need to bring in elements that were out of step with the workshops but better coupled with the activities of the community, that the name "Distillery" emerged, as these workshops were used to distil the ideas that came out of the plenary sessions. The latter are now prepared in such a way as to think about the Distillery, the plenary sessions thus becoming the Distillery's working subjects. In this sense, we can say that the plenary is today in the Distillery since it is inserted in a cycle of three phases beginning and ending with a Distillery that frames the plenary. The aim of the Distillery is to produce deliverables to enable participants to continue the reflection initiated outside its walls, and it is with Brigitte Romagne, a specialist in perfumes, that we will imagine the form that these deliverables will take today. There are three types of perfume: head, body and trail, each corresponding to a different evolution of the fragrance. Similarly, there are three types of deliverables based on this typology:

— In the short term: these are "tips and tricks" that help answer the question on "what could be operational right away?"
— In the medium term: these are ideas and concepts that are accessible but difficult to implement quickly. For example, one of the emerging ideas was a chess room inserted in all companies whose goal would be to make failure a teaching experience.
— In the long term: to address deeper issues and understand what new questions they may generate. In this sense, we are no longer in the immediacy.

The outcome of the Distillery is the magazine of the *Unpossible*. The latter is an integration of all the productions that take place during the plenaries, the Distillery or during events outside these two meetings. Restituted in the form of articles, new words, games, bibliographies, drawings, models, representations, quotes from authors, the magazine of the *Unpossible* is the image of the Distillery, a multiform conglomerate, a rich and fertile "travel diary".

— *The Boxing*
The *Boxing* is an element that is increasingly found outside the boundaries of the community. It is the idea of setting up a sequence that will allow a project leader who is in the process of developing a project to benefit from a 15-minute presentation followed by an organized return of the 80 people from the plenary with a separation of the room into two parts: knights and dragons. The first part highlights the positive elements that would ensure the success of the project and the others point out more the brakes, risks and obstacles to its implementation. This type of situation is clearly linked to the DNA of the community, which is based on fertile friction and "dispute". Thus, the positive points that the project leader has not thought of can finally be integrated and the negative points that are risks are turned into positives so that the project is enriched. The boxing session ends with a written synthesis of the points raised by the community. The project leaders leave with something concrete. In this sense, it is a good way to present and spread projects to put them in direct confrontation with innovation experts.

Guillaume Tilquin, project manager at Renault, has twice used the *Boxing* from the Innovation Community to push forward the concept of a platform for new electric commercial vehicles coupled with a new ecosystem for Renault.

3. *The trace*

The question of the trace that the innovation community must leave remains complex both in terms of its supposed necessity and the form it will take: summary report, recommendation, etc. All these ideas have been discarded because they would lock in certainty, whereas the essence of the community is to remain open to the gap. Several formats have been tested: in visual form refocused on the emotional, but this is not enough and not always very clear.

Today, a website has emerged for the community members with reports in the form of restitutions, videos of speakers, supports of presentations of the actors. The goal is to build a "trace" in the form of a very simple "exploration notebook". Many social networks already exist and mobilize a large part of the users' attention. In this sense, there was no point in adding one more. The website is on a page with a timeline. One walks around the timeline with a cursor through a chronological filter. In a few clicks, a member of the community can find a workshop, an intervention, a presentation, bibliographical elements…

Laurie DA SILVA's creation of scribes (comic strips made during the sessions) during the plenary sessions and workshops also contribute greatly to capitalizing on the exchanges in visual and pictorial form.

VI. Obsolescence

1. *Rituals that freeze, regenerate participants and formats, place*

Renault's innovation community is a living thing and like any living body, the question of its possible death must be asked. For the founding members, this possibility was envisaged from its birth. Unlike other innovation communities, Renault's innovation community should not be an object frozen in time, yet during its nearly ten years of existence, many rituals seen at first as innovative now seem to constrain it and prevent its development. Yet innovation is a constantly moving organism that needs renewal to realize its potential. Anne Bion, who worked at Renault and is now Director of Innovation at Nutriset, a food company, is a good observer of the community thanks to this dual internal/external status. As a result, she wonders about the connection between Renault's innovation community and smaller companies around the issue of mobility: "I find it difficult to find my way around now because this community deals with a particular theme and organizational problems that are no longer mine because the company I joined is of intermediate size and deals with agri-food issues. While the exploratory dynamics and the innovation projects are always very interesting, the subject matter and the change of scale are very different. My field of action today is linked to the countries of the South, the antipodes of Renault's concerns. Imaginations are therefore opposed". It is therefore legitimate to question the relevance of the community's contributions for all its members, but also how to introduce renewal. If we are convinced that this renewal will come through the regeneration of the internal dynamic, we wonder about the possibilities of this regeneration. Is it necessary to change the format of the plenary sessions? Restructure the rituals? Introduce new object-terms?

2. *"Enlightened" desires*

More and more companies are following Renault's example and setting up their own innovation community, which underlines the growing interest in this type of structure, and consequently the relevance of its creation. These new organizations can take different forms: Open Lab, Fab Lab, I-lab ... depending on the needs and topics addressed by the major groups. This competition is welcome because it proves the need to think differently and it allows a wider range of subjects to be addressed. However, it is important that the multiplicity of communities does not lead to a dispersion of resources between the different communities. From this limit arise frictions on questions of fairness and balance between the different communities that participate in what could be called ethics. Indeed, this risk may be all the greater if the community created requests financial participation to join it. As a result, the members of this new third place will have more willingness to refocus on the events of this Open Lab to capitalize on their initial investment.

Such an attitude could then prevent the innovation network (all the innovation communities together) from being set up, all the more so if the company at the head of the community decides to no longer participate in events outside its own. Going "in isolation" would be a counter-productive attitude and contrary to the innovative philosophy of introducing the collective to create the confrontation it needs. It would mean moving from a systemic approach that makes sense with innovation to a more individual approach. Thus, creating new communities is essential, but this can only be useful if we establish a link with the outside world, particularly by participating in events in other communities.

3. *The return to DNA as a condition of existence*

The innovation community has been built on the interest of reconciling innovation and rationalization and it is clear that more and more initiatives initiated by other companies are moving in this direction. Indeed, these other companies, which may face the same problems as Renault, are choosing to change their stance and create differently. This change of culture will be a sign that the mission of the innovation community will be completed and successful. If the goal has not yet been fully achieved, the evolution is underway and it also raises the question of what comes after. Indeed, there is today a real willingness, expressed by the members as well as by the founders, to challenge the innovation community both in substance and in form. The aim is to see whether it not only resists the challenges it faces but also whether it is capable of regenerating itself. These challenges are the same ones that motivated its creation and which it seeks to combat on a daily basis by injecting renewal and rupture. Indeed, today the innovation community aims to avoid falling into the traps of habit that could jeopardize its raison d'être. For, it is appropriate to ask what will become of the community once it has met the expectations of its members. Of course, innovation cannot come to an end, but each year the community asks us about the conditions for its renewal. For the moment, we do not yet have an answer, but we know that it will not be able to do without what makes up the DNA of our community. This DNA is the Distellery, the boxing and the friction in the plenary which will immediately produce reflections of fertile doubt. These three elements have in fact in common to emphasize the dispute and the tensions in order to create a richer reflection, the first step necessary to create innovative ideas, concepts and projects.

4. *Taking one's place in the ecosystems on the move*

Taking one's place in moving ecosystems is a sign of a willingness to remain alert to how one can transform oneself. The important thing is to be able to fit in at the right level without claiming authority. These ecosystems are multiple and multifaceted since they can be third-party, open lab, FabLab or multi-partnership. They have all been created to respond to innovative projects that make sense with those of the community and that is why it is important to find a place for them as well.

VII. In Conclusion, the Philosopher's Reflection

We have asked Dominique Christian, the philosopher, to conclude this chapter with a quirky opinion as he does in his interventions during the community plenary sessions.

"My first reaction to this chronicle of Dominique and Fred will be a reference to Tchouang Tzu. (It will be understood that I am trying to illustrate here the method used in the session, that of a *desachant*, i.e. someone who "removes what he knows"[5]). In a text by Chuang, Confucius confides to the novice: "I am trying to wander in my words, are you ready to wander in your listening?". What [the word] requires above all is not adhesion or approval, but a certain quality of listening or, more precisely, a certain interior disposition that commits us to the exchanges inscribed in the words. Tchouang Tseu makes Confucius speak, who specifies it: "I can proceed [that is to say, instruct you] only if you are 'without anything' [that is to say, empty of ideas or ready-made theories]". And the device, the operating protocol proposed by the community, is first of all this invitation to emptiness, a theme that we have come across in our meetings in the past. To do this, the visit of a sage[6] must plunge the interlocutor into the throes of perplexity. The intervention is effective, it produces a revolution in the "patient", his head is turned over and he no longer knows, in terms of conversion, to which saint to dedicate himself. It is not an easy task. According to Jean François Billeter, the great translator of Tchouang Tseu, it resembles hypnosis, but unlike it, it does not provide any appeasement, it does not heal from any complex; it has no intention for the other.

And this leads me to a second remark, which concerns the way Dominique Levent masks her role in this text, as in the plenary session. For there must be conditions for a collective to take the path of inscience.[7] This presupposes a great deal of confidence in the interest of a system which has the declared intention of not producing more knowledge. This presupposes a great deal of confidence on the part of each individual. Of course, the presence of highly experienced people in the field counts, of

[5]The one who removes what he knows is to the one who knows what the stain remover is to the stain. It aims at what Jankelevitch called nescience, the non-knowledge that allows us to grasp the je-ne-sais-quoi that always escapes knowledge and that distinguishes the craftsman from the artist. Disruptive innovators can only be artists, whereas methods for innovation are the tools of artisans.

[6]In China, there were wise men, who possessed wisdom; in Europe, we are only philosophers, lovers of a wisdom to which we aspire.

[7]The word processor made by Apple engineers systematically corrects inscience into insolence. I mean inscience, even if it bothers the producers of computerized pseudo-knowledge.

course, the presence and the often-pointed interventions of academics are essential, but it seems to me that it is Dominique, as a fragile master of ceremonies, who contributes most to this. She constantly makes visible her insecurity and the work she does to put up with this discomfort. This gathering of innovators in vulnerability and doubt may be like a session of Alcoholics Anonymous, and there is nothing pejorative about it: it is the only method that really works for detoxification. Even if the participants have to sadly identify themselves as alcoholics though they are sober.

For innovation is detoxification. For a long time, it was thought that it was routine thought and practice that the innovator had to emancipate himself from. Today it seems to me that innovation must fight not, or not only, against routine, but also and above all against commercialization and entertainment. The classic criticism of change is no longer "it's not: it's impossible", but "it's not profitable". Hatchuel's preoccupation with the theme is an important element and we note the passage in his work, which has a strong influence on many participants, from a preoccupation with methods (C-K…) to an investment in the nature of the enterprise and the relationship between wealth production and financialization. So can the community last, should it last? It is difficult to clarify the extreme utility of the useless to managers in a hurry and boarded by finance. It is the lot of the whole world, companies, states, care services today. Reason does not always prevail, and many worlds have disappeared under the deleterious effects of King Money. But as for the proposed system of common distancing, it has been working for 2300 years, it should still be useful for a few more years, while companies regain their raison d'être and humans regain their pride and free thinking".

List of Renault Innovation Community Members

List of community member companies:

La Poste, Renault, Altran, Dassault Systemes, Thales, Alenia, Orange, Visteon, GDF Suez, Without Model, Syb Consulting, Kopilot, Sita Recyclage, Qwant, Centre Michel Serre, Bouygues, CEA, Agence Babel, SNCF, Bosch, Perwit, Fing, Salomon, Shell, La Coentreprise, Air Liquide, EDF, Desdoigts Design, Rhodia, Valeo, Biomerieux, Sismo, Airbus, Safran, Michelin, Gruau, Total, Eranos, CEA Minatec Idea's Lab, Renault Design, BNP Paribas, Sosciences, Chanel, Schneider Electric, Auchan, Essilor, Merkapt, Ouishare, Near Future Laboratory, Dim, Saint Gobain,

Kisskissbankbank, Usbek et Rica, L'Oreal, Groupe SEB, Danone, Renault DVY, Centaury.

List of schools that are members of the community:
Pôle Universitaire De Vinci, Polytechnique, ENPC, HEC, CNAM, École des Mines, ENSCI, Université Grenoble, Collège Polytechnique, Strate Collège, École de Management de Lyon, Collège de France.

List of speakers who are members of the community:
Cognitive psychologists, Consultants, Photographers, Philosophers, Designers, Writers, etc.

© 2021 World Scientific Publishing Company
https://doi.org/10.1142/9789811234286_0008

Chapter 8

"Ecosystem of Innovation", A New and Efficient Business Practice: The Case of the Open Lab Michelin

Erik Grab

When referring to the different waves of innovation that have taken place in companies over the last few decades, the most recent one, "Ecosystem of innovation", promises to be the one that will have the greatest impact on the societal demand that companies have to meet.

The wave of "Ecosystem of innovation", by taking better account of the expectations of consumers and citizens, by seeking to jointly optimize the contributions of each actor in the process of change (companies, academics, startups, public authorities, professional sectors, international organizations, etc.), and by distributing the value created between these different actors equitably, will revolutionize our internal working methods and our external collaboration methods.

"Ecosystem of innovation" will require new academic training and new business practices. It is from this perspective that the Open Lab Challenge Bibendum experience is retraced in this chapter. After a review of the most recent waves of innovation that the Michelin Group has experienced, and a brief history of Challenge Bibendum, which is at the origin of the "Open Lab" ecosystem, this chapter seeks to explain the mission and principles underlying this ecosystem, as well as its main operating methods and daily practices, to lead to a reflection on the future of the Open Lab's innovation communities, which are focused on Sustainable Mobility.

In fact, an innovative ecosystem, in order to function properly, should be inspired by the great philosophies, whether ancient, modern, post-modern or contemporary. Both are defined by three components:

— The "theoria" or the intelligence of what is. Within the Open Lab, members of the innovation communities call this "shared diagnosis". Focused on the issues of transporting people and goods on a global scale, the Open Lab ecosystem first sought to describe the "playing field" and to share a diagnosis of the issues and challenges to be taken up in terms of sustainable mobility at the global level. Hence, its closer ties in recent years with major international organizations such as the UNO, the ITF (International Transport Forum), the IEA (International Energy Agency), the World Bank and the WBCSD (World Business Council for Sustainable Development), with which the Open Lab now works regularly.

— Ethics, practice and the thirst for justice. It was not only a question of sharing on a global and somewhat theoretical level the "mobility playground" of the Open Lab's respective companies, but also, on each theme given and prioritized in common, of defining the roles of each company member of the ecosystem, its potential contributions in the various communities of interest formed, and its level of involvement in the "Corporate Advisory Board" (which is examined later in this chapter). In short, it was a matter of defining the "rules of the game" between the different companies participating in the ecosystem, particularly with regard to the creation and distribution of the potential value created by joint initiatives. It is a question of respect for each of the members, and therefore of a certain morality, an ethic to be applied on a daily basis, in order to ensure the sustainability of the ecosystem.

— Finally, and perhaps most importantly, it is necessary to define a form of wisdom and the quest for salvation … as in Plato or Kant! … The question is why do companies agree to come together as an ecosystem, beyond the partisan interests and the pursuit of growth and profit of each respective company participating in the Open Lab? What is the purpose of this ecosystem, what is its ultimate and higher goal? And finally, why do these companies make all these efforts to get to know each other better and to work together better? Do they share a common vision of what needs to be done to make mobility more sustainable and better? What is ultimately the real *raison d'être* of the

Open Lab Challenge Bibendum? These questions were strongly debated within the Open Lab ecosystem when it was created and are still systematically raised today when new companies join the Open Lab. They actually bring back the real role of the company to address societal issues and consequently create value for its shareholders and customers and generate salaries and fulfilment for its employees. It is in this order that every company must act, in a quest for a higher role that goes beyond the simple search for growth and profit. It is with this higher mission in mind that the company transcends itself. For an ecosystem, a shared higher mission, coupled with the necessary ethics and mutual trust of the people involved, will make it possible to overcome the fears that co-innovation and *a fortiori* open innovation inevitably raise within our companies and to reach a certain form of "wisdom" with regard to innovation.

I. The "Waves of Innovation" that are Shaking Companies Up

Michelin, like many other companies, has experienced various waves of innovation that have passed through the group's teams and subsidiaries around the world. These waves gradually led our small Open Lab team to realize that ecosystem innovation and its corollary, communities of innovation or "communities of interest" as they are known within the Open Lab, were probably the only true answers to the challenges that the world of mobility would have to face in the coming decades.

The first wave in the 1980s was the "global innovation" one and the question of whether or not Michelin should decentralize its R&D centres in countries such as China or India, with what degree of involvement of local teams, and while taking into account the multiple challenges of protecting innovations, patents and "house" secrets.

Then, in the 1990s, the question of the "co-creation with consumers" in the sense of the academic authors C.K. Prahalad or von Hippel was raised. This wave of innovation has notably encouraged Michelin to get closer to its major road transport, mining or airline customers, for example, and to offer them "PSS" for "Product, Services System", in other words, sales of kilometers to road transport fleets, tons transported to mines, or landings or takeoffs to airlines rather than tires. This wave of functionality savings within Michelin has greatly helped the Group to put the customer even more at the centre of its activities. For example, some members of our "Michelin Solutions" sales force have gradually become consultants to transport companies seeking to optimize their customers' business models and profit and loss accounts. When we offer driver training for more responsible, safe and energy-efficient driving, and when we use our ultra-low rolling resistance tires to help our trucking customers save fuel, we are addressing the two main cost items of our customers. Similarly, when we transform our tire distribution network into a network of service providers, we rely on an intimate knowledge of our customers' needs.

The next wave of "open innovation" in the early 2000s was much more complex to integrate, with strong internal reluctance to open the group's research centres to the outside world and to include the latter in Michelin's R&D projects, even though cooperation with universities is currently being stepped up. It was on this occasion that the question began to be raised as to the degree of openness of what would later become the "Open Lab" (see Section IV).

Then, at the beginning of this decade, the group sent teams of Westerners to rub shoulders with Indian frugal innovation methods. This new wave of innovation prompted us to go and design so-called "affordable" tires in emerging countries and for emerging markets. The idea was to impose strong constraints on our teams: use of existing industrial equipment, a very significant drop in manufacturing costs, limiting the quantities of raw materials used per tire produced, etc… in order to make them think differently and frugally throughout the value chain: from raw materials to distribution, including costs outside Michelin's internal processes. The objective was also to improve the "time to market" allowing a market maturity in 3 years in industrial operation, where 5 years was the norm. To achieve this, we had to think outside the box in terms of methodology and call on the creative diversity of our teams around the world: North Americans, Indians, French and a Mexican were mobilized and based in India.

The cost reduction achieved and, more generally, the success of the new frugal offer were such that we asked this innovative team to train our Western industrialists in particular in these new and more efficient methodologies and manufacturing methods. This was the wave of "reverse innovation" developed by C. Trimble and Navi Rajjou (who is one of the academics who knows and appreciates the Open Lab Challenge Bibendum ecosystem).

These different waves of innovation have accumulated in terms of internal experiences and learning and now allow the group to integrate with serenity, and for several years now, the latest wave of innovation to date: "innovation in ecosystem". It is important to note at this point that for Michelin, the term "ecosystem innovation" in no way refers to the concept of "extended enterprise" that many of the Open Lab's partner companies have been practising for some time. The "extended" company calls on its "traditional" environment: its suppliers, its business partners, such as its distributors, for example, and of course its customers, whom it may also bring together in innovation communities that include, for example, informed consumers (lead-users) and innovation teams (marketers, R&D, etc.) in order to develop new products or services that are closer to market needs.

Examples of extended businesses include the case of Toyota and its network of preferred suppliers with whom the links are so close and intense that Toyota's managers do not hesitate to say that "the focus of innovation for the group is the network, not the individual firm" (cf. Dyer and Nobeoka, 2000). Another example is that of Procter & Gamble, which after its change of strategy at the beginning of the 2000s to favour an open innovation strategy, made a point of stressing that the group's innovation forces now included buyers, consumers and users. Thus, to describe the company's research strengths after the adoption of an open innovation model where consumers are now invited to transmit their creative ideas, the group's CEO did not hesitate to state that "before (moving to an open innovation model), we could count on 7,500 researchers, whereas today we can count on 7,500 +1,500,000 researchers".

The ecosystem-based innovation approach that we are currently developing at Michelin is in fact much broader than the extended enterprise approach, and in particular involves entities that have often never had to deal with our group's subsidiaries, at least not on the innovation subject in question; but these entities potentially have technological building blocks or business models that could provide all or part of a mobility solution to be developed in the future. This is the case, for example, when Solvay, Dassault System and Michelin work together on exoskeletons and seek out, within a community of interest, startups and academics who can join them to both better understand the future demand for exoskeletons and build initial prototypes to validate this demand during experiments in factories or with the general public.

The wave "Ecosystem of innovation" therefore seems to be a more ambitious one, but also a more pragmatic and efficient one, as it seeks skills and resources where they are most accessible or cheapest. In other words, it seeks to optimize the sum of the contributions of each of the entities and to make the overall output of the community as efficient as possible at a given stage of innovation. But to do this, as the case of Michelin's Open Lab shows, you need to benefit from a large, diversified and international ecosystem that is not built overnight.

II. A Brief History of the "Challenge Bibendum"

It was in 1998 that Edouard Michelin and some of his close collaborators conceived the Challenge Bibendum. The pretext was to celebrate the birthday and 100 years of the Bibendum or "Michelin Man" in North America. But the idea of creating an international ecosystem to address major technological challenges and business models and to develop more sustainable mobility was already germinating at the end of this century.

The first Challenge Bibendum events in the 2000s focused on new technologies and gradually brought together the major vehicle manufacturers, their equipment suppliers and the world's major energy producers. Numerous prototypes demonstrated that tomorrow's mobility would be multi-technological and in harmony with the environment.

The 10th Challenge Bibendum in Rio de Janeiro in 2010 marked an important turning point when President Lula, who introduced the event, called on Michelin's ecosystem to make tomorrow's mobility not only "safer, cleaner and connected", three of the Group's key words until then, but also "accessible and affordable" with a view to the millions of poor people in the favelas or shantytowns who do not have access to decent mobility. Greater consideration of the famous "base of the pyramid", which still represents billions of disadvantaged people around the world, has, for example, led the Open Lab team to develop an inclusive mobility approach with Total.

Then it was Berlin where the number of participating entities has considerably increased: public authorities, international organizations, cities, startups, universities ... have joined the efforts of Michelin and its partners to take together different concrete initiatives and produce different publications in favour of sustainable mobility.

During the Challenge Bibendum in Chengdu (China) in 2014, Michelin and its partners moved to a "Think and Do Tank" stage. To take two very different illustrative examples, the Group and its partners have conducted experiments on new technologies such as autonomous vehicles in "real life" with research into use cases (the so-called "Design Thinking" approach), or cooperation with the UNO at COP 21 in Paris and COP 22 in Marrakech at the end of 2016, in order to strongly involve the transport world in reducing greenhouse gas emissions and limiting global warming.

In this last example, the importance of a structured ecosystem that has developed a shared vision of the mobility challenges and solutions to be provided at the global level is crucial. This is what attracted the United Nations and the Ban Ki-moon teams who offered us the opportunity to be their Transport Partner.

Until then, and as surprising as it may seem, transport was not included in the climate negotiations. As of COP 21, in collaboration with diverse UNO partners such as SLoCaT (Partnership on Sustainable, Low Carbon Transport), Investment Banks, NGOs, etc., the Open Lab Challenge Bibendum developed a new process "on mobility and climate" that should lead to a drastic reduction of greenhouse gas emissions from the world of transport and thus make its contribution to limiting global warming to less than 2° (reference: IEA scenarios). It is also in Chengdu that telephone operators, international companies or startups from the information and communication technologies, banks and investment funds joined the companies involved in the Open Lab, thus expanding the "mobility" ecosystem.

The Open Lab team now has a sufficiently large, international and diversified ecosystem to "surf" on the "ecosystem innovation" wave.

III. The Mission and Principles of Michelin's Open Lab Ecosystem

At the end of the Chengdu event, which brought together more than 5,000 professionals and more than 500 journalists who appreciated the "experiential" workshops set up by companies as diverse as SAP, Mc KINSEY or IBM, the conferences, including the one by TED which had staged its first TED Talks in China within our event, the "Ride & Drive" of various prototype vehicles, etc… a dissatisfaction remained, including within our team.

A verbatim statement by FAURECIA expresses this latent dissatisfaction: *Challenge Bibendum is truly fantastic! For a week we are rebuilding the world of mobility together! ... But then nothing more! You have to wait two years to find yourself in such a context. In fact, not much happens between each Challenge Bibendum event ...*

In other words, each Open Lab Community of Interest first goes through the "strategic anticipation" phase of the evolution of the environment before being able to claim to innovate together or influence the outside world. The 3 pillars of activity of the Open Lab are thus in order:

(1) Strategic anticipation.
(2) Co-innovation.
(3) The influence, in particular of public authorities (regulations, standards, etc. ….).

We can say that our communities, which are initially formed around a "leader" (I come back to this in the next paragraph) and an interest in a sustainable mobility challenge, are gradually becoming, if all goes well, communities of innovation. Because what could be more normal than wanting to co-innovate once you share the same vision of the problems to be solved and the solutions to be provided!

Since the main mission of the Open Lab Challenge Bibendum was to develop more sustainable mobility on a global scale, it could not be a question of dealing with subjects of interest to only one or two members of the Lab's participating communities. It could not be a question of making these communities substitutes for innovation or marketing thinking that is sometimes insufficiently developed within the respective organizations. On the contrary, the challenge was to deal with the common good of clean, safe, connected and accessible mobility throughout the world.

At the same time, however, legitimate privacy issues had to be managed: how far were we allowed to share the playing field, market diagnostics and technology roadmaps? How could we prevent more or less direct competitors, who are nevertheless members of the Challenge Bibendum ecosystem, from joining the so-called "sensitive" communities of interest as "stowaways"? And, even if we limit ourselves to sharing our diagnosis, doesn't that already reveal part of our strategy?

All these questions led us to quickly adopt a founding and extremely structuring principle: each community of interest would be headed by a "leader" company that would determine all the conditions of its operation: its mission, its precise objectives (see the community launch sheet in Appendix 1), its participants that it would co-opt as it wished, its budget, the number of its physical meetings, its roadmap… but also what could or could not be shared by the community with the outside world, including within the Challenge Bibendum ecosystem itself.

Thus, the leading company, usually also a member of the Open Lab's governance body, is all-powerful within its community, and is in charge of determining what it wants to share or not with the other invited companies or organizations, but knowing that it will not be able to attract "useful" participating companies to its community if it is not willing to go relatively far in sharing. Before receiving you must first give!

It is with such a very simple operating principle, which contrary to what the term "Open Lab" might suggest is quite far from a total opening to the outside world, that our Lab can afford to have within its Corporate Advisory Board four energy specialists (Air Liquide, EDF, ENGIE and Total) who live within the ecosystem in good intelligence. On some subjects, these members share with their colleagues, who are nevertheless also competitors, but on others, they do not invite them into their communities, and this is perfectly well experienced by everyone. It is even conceivable that each of them can create a community of interest on the same subject, obviously with different participants, without this in any way disrupting the smooth functioning of this approach.

After the Challenge Bibendum event in 2017 in Montreal, the members of our Corporate Advisory Board begun to build new communities to prepare and feed into the conferences and workshops at the event. Challenges such as the transformation of passenger transportation under the dual influence of vehicle automation and collaborative mobility or the new actors and business models of urban logistics are being analyzed.

With experience, the question of leadership has proved to be key and the adopted approach has considerably reassured the general management who now agree to send their best experts to the communities. The consulting agencies themselves understand that it is in their interest to invite their clients and prospects to the communities in which they are leaders and to share with them the first elements of diagnosis, even if it means proposing to go further with a pooled budget that is usually theirs naturally. This is how Cap Gemini or Oliver Wyman, for example, have created communities of interest, respectively, on the impact of the Internet of Things on new mobilities and on the new business models of last mile delivery.

To date, communities have managed to overcome the particular interests of their members quite easily, especially as they often find that this allows them to deal indirectly with the latter.... Indeed, there is no incompatibility between the search for new markets or new areas of opportunity for businesses and the pursuit of the common good. More often than not, there is convergence between meeting the needs of customers and taking into account the expectations of citizens.

This being the case, we also sometimes have failures! More often in the way we animate the community and more often in the form than in the substance. I'll take an example about a research project on the "Automobile Competition of the Day After Tomorrow", a community of interest under the joint leadership of ACO (Automobile Club de l'Ouest, organizer of the 24 Hours of Le Mans) and Michelin. For the "Kick Off meeting", we had gathered together a group of "gentlemen drivers" company managers, former F1 and rally drivers, car event organizers, … and we had presented them our collaborative web platform. All of them had sworn to us that, back home, they would start, from their personal computers, to post articles, comments, suggestions, ideas about the car racing of the future… A few months later, we were still waiting for these promised contributions!… We then decided to go back to the good old qualitative and quantitative questionnaire sent by e-mail to each participant as well as to physical appointments around a good table! And the contributions did not wait any longer!…

The lesson of this failure is twofold: collaborative tools on the web are not always adapted and practised today by all categories of the population and we have to take this into account from now on. Physical appointments must be mixed with virtual appointments for a community to function optimally.

Thus, the Open Lab ecosystem and its permanent collaborative platform are today participating in the development of common positions in the world of passenger and goods transport, in the anticipation of tomorrow's mobility markets, in the acceleration of technological innovations or business models, in the launch of market experiments and joint initiatives, and in fine-tuning the invention of better mobility for people and for the planet.

IV. How the Open Lab Communities Work and Practice on a Daily Basis

The main operating principles of the Open Lab communities that have just been mentioned are not without a certain number of rules of the game that have been progressively identified and disseminated within the ecosystem in a pragmatic manner.

If we begin by mentioning the deliverables of these communities of interest, they can be of different natures: In some cases the members wish above all to develop together a few plausible development scenarios, starting from the idea that if these scenarios are shared within the community, which, if the recruitment of the leader has been relevant, includes representatives of the main stakeholders in the subject, they have every chance of being implemented and therefore of becoming reference scenarios for the entire profession. Other communities, having observed that they were working well, that mutual trust had been established between its participants and that they shared the same diagnosis of the challenge to be addressed, decided to go beyond a prospective vision and undertook to build together offers that potentially met the future demand identified. This is how the Cycles Lapierre and the Dutch group Accell developed within their community a prototype electric urban vehicle aimed at ensuring a soft mobility from 7 to 77 years old.

Other communities go even further and simply decide to co-innovate, produce and develop new mobility products and services. For example, Robosoft, a company specializing in robotics, and Ligier, a manufacturer of license-free vehicles, have created a joint venture called Easymile to produce and market a driverless vehicle that has already been acquired by several companies and cities around the world. In the latter case, the question of the fair distribution of the value created is crucial. The Open Lab encourages such communities to quickly build a macro economic model, based on a canvas by Osterwalder and Pigneur (2010), for example, so that they can establish value distribution keys well in advance of the breakthrough innovation to be produced. Even if the economic model is generally false at this stage of innovation maturity, its very existence from the very beginning of the community is likely to reassure the participating companies and create a climate of mutual trust.

Finally, it should be noted that after only few years of existence, the Open Lab is already generating "daughter communities". Thus, a new community of interest, concerning new models of management of urban space and parking lots, has emerged under the leadership of E&Y, starting from the above-mentioned mother community and including almost all of its members. If the participants feel their best in an efficient and trusting working environment, with clear rules of the game and applied ethics, why not focus on foresight and ecosystem innovation!

But in order to achieve such deliverables, the Open Lab team quickly realized that a process as well as several proven methods and tools had to be implemented. Working efficiently in an ecosystem does not make sense! Each of the Lab's partner companies generally already knew how to work directly with a startup, a professional organization, a public authority or a city. But working at the same time with all these actors having different cultures, processes, methods and innovation tools was a real challenge for our communities. Their participants therefore had to open up to others, take the time to understand their innovation constraints and learn about the new common methodologies proposed by the Open Lab. The Open Lab team recognizes, for example, that without the unconditional support and the conceptual and methodological input of personalities such as Michel Godet, former Professor at the Conservatoire des arts et métiers and member of the French Academy of Technology, or François Bourse, who is both a teacher and consultant at Futuribles, the ecosystem would probably not have progressed as quickly within its communities of interest, which today use common tools that they enrich on a daily basis.

As a concrete example, we can take the Lab's prospective web (or "Prospective Web"), which aims to summarize in a single diagram a future multi-dimensional space of innovation. This web was originally designed in partnership with the company Visteon, then enriched by practice by Michelin and recently further improved by Solvay (see Appendix 2 and an illustration on urban mobility).

The Lab's team was therefore led to develop a toolbox with the help of academic personalities but also various consultants including Mc Kinsey, Oliver Wyman, Roland Berger, At Kearney, Arthur D Little, Accenture and Futuribles. With Futuribles, the Lab is developing training in foresight and innovation. It will be supplemented by MOOCs (Massive Open On-line Courses), the first of which was created thanks to Michel Godet, which will be offered to the employees of our partner companies. The idea is to facilitate both training in the methods and tools of the Open Lab and initiation into the various facets of sustainable mobility, in order to raise awareness of the challenges to be met in our sector of activity and to facilitate everyone's participation and creativity in future communities of interest. The Open Lab team is therefore developing a bank of MOOCs with the partner Coorpacademy founded by the former CEO of Google in order to allow a massive, interactive and playful deployment of some of the contents of our ecosystem. For example, Air Liquide will contribute to this bank by providing two MOOCs, one on hydrogen and the other on biomass.

This toolbox, these trainings, these MOOCs obviously have a cost. This is why membership of the Open Lab is not free. This membership also covers the many workshops and conferences that we organize during the year on subjects of common interest. For example, in 2016, the Open Lab organized working sessions on mobility topics in such diverse configurations as incubators or startup accelerators, with the European Commission or with the cities of Copenhagen and Göteborg.

Finally, in this summary chapter on the practices implemented, we should mention the importance of the Corporate Advisory Board (CAB). It is the CAB that decides on all of the Open Lab's activities, events, investments and annual priorities. Its members are all representatives of major companies involved in the world of mobility, such as Thales, the latest to join the CAB, or DHL, Geodis and CGI who are preparing to join. In order to avoid a plethora of members, which would make governance difficult, the Lab has invited professional organizations. Thus, for example, the French automotive industry, made up of more than 5,000 companies including Renault, PSA, Valeo, Plastic Omnium, Faurecia, but also a large number of small and medium-sized subcontracting companies, is represented by the PFA (plateforme automobile), its professional organization.

Sharing the same methods of strategic anticipation and innovation, the same vocabulary and the same practices is necessary to develop shared diagnostics, an intelligible vision of the playing field and to act together in communities of interest. But this is not enough to ensure the sustainability of such a vast and diverse ecosystem. It must be endowed with a societal purpose, a higher goal that transcends it and that brings in its wake the considerable strengths and capacities for action of its various stakeholders.

V. "Sustainable Mobility for All", the Raison d'être of the Open Lab, and its Ultimate and Superior Goal

From the very beginning of its reflection on the opportunity to create a forward-looking and innovative ecosystem in the field of sustainable mobility, my team has sought to establish a platform of knowledge and action levers that would be of interest to as many of its stakeholders as possible. From this point of view, the "Green Paper" published for the last Challenge Bibendum and co-authored with Patrick Oliva, one of the "fathers" of the latter, has made its mark at the international level by formalizing a "shared vision" by all the members of the ecosystem of what sustainable mobility should be at the global level and by providing solutions or "Game Changers" that have since been adopted by a number of international organizations, public authorities and corporate partners to guide their respective actions.

Indeed, to mention only this last challenge, without inclusive mobility, i.e. without safe and accessible mobility for the billions of poor people spread over all continents, there will be no access to education, health care and employment, and therefore, in the long term, no contribution to the economic development of the countries concerned.

It is the in-depth analysis and recognition of these societal challenges by all the partners of the Open Lab, as well as the awareness of the impact that we collectively could have on each of them, that has enabled us to transcend ourselves as an ecosystem. This "shared vision" led us to prioritize five breakthrough initiatives on which we continue to work today within the various communities of interest. Some of them will require unprecedented innovation efforts on the part of our companies and strong strategic choices towards the development of more sustainable mobility. But the momentum is on and it will not stop. This is one of the many advantages of working in an ecosystem: the ripple effect and the stimulation of all the participating organizations in view of the initiatives taken and shared by each of the members. This "shared vision" transcribed in this Green Paper therefore still today represents a reference base for all the communities of interest of the Open Lab Challenge Bibendum ecosystem.

However, my team is already preparing a new collaborative and updated vision of this book with one of the Lab's partners, BLUENOVE and its collaborative web-based tools, because the changes in mobility that everyone is facing today are accelerating and the Open Lab team wants to integrate into the ecosystem new players who have already contacted it, such as UBER or BLA BLA CAR. But we believe it is essential that these new players first adhere to the Open Lab's *raison d'être* and values.

Appendix 1: Community profile at the Open Lab Michelin

Michelin Challenge Bibendum Community
A Think and Action tank for sustainable mobility

COMMUNITY PROFILE

First Name	Name	e-mail	Phone	Title	Company

First Name	Name	e-mail	Phone	Title	Company

Members

N°	First Name	Name	e-mail	Title	Company
1					
2					
3					

The community

Community Name (50 characters max including spaces in english)	
Description of the community (250 characters including spaces in english)	

Objective of the community (identified topic, goal…)	
Expected deliverables Eg: position paper, benchmark, green paper, strategic anticipation scenarios…	
Privacy level	○ **Open Group:** Every user can freely read and create content ○ **Membership only:** Every user can view content, but they need to join the group to create or comment ○ **Private Groups:** The Group is visible, but content is not visible to non-members. Users of the system need to request membership to join the Group or need to be invited. Group managers need to approve these requests ○ **Secret Group:** Secret Groups are not visible to non-members. Members need to be specifically invited into the Group and Group managers need to approve their membership
Key dates (event, meeting, official presentation, deadline…)	
Animation language	◉ Français ○ Anglais ○ Other

Comments	

Appendix 2: The Open Lab Michelin, source É.Grab.

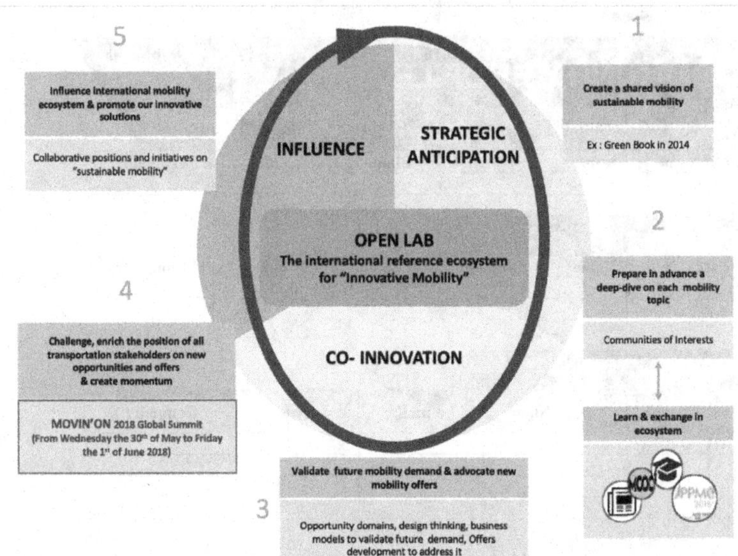

Part 4

Mobilizing Communities of Innovation Through the Middleground

© 2021 World Scientific Publishing Company
https://doi.org/10.1142/9789811234286_0009

Chapter 9

Innovate with an Online User Community

Guy Parmentier

More and more firms rely on online user or brand communities to design new products and services, and to enrich them with new content and features (Burger-Helmchen and Cohendet, 2011; Jeppesen and Frederiksen, 2006). These companies, such as Lego, SNCF, Iliad, Orange, Decathlon and Nadeo, use digital technologies to communicate with users of their products and services and offer them content design and creation activities. Communities that regularly gather and connect users of a product and service on the Internet become a new source of innovation and enable companies to renew their innovation capacities.

However, though this principle seems attractive in itself, its implementation presents many challenges. The creation of a community cannot be decreed, and it is not inevitable that users are ready to contribute or that they have good ideas. Sometimes users' ideas are interesting, innovative and value-creating, but not aligned with the firm's strategy; sometimes their ideas are difficult or even impossible to develop. In addition, a community tends to run out of steam quickly and it is very difficult to maintain interest and participation over the long term.

Co-creation practices with communities of users have developed mainly since the 2000s and by observing some of the pioneers, we can draw lessons from their experience. Nadeo, who developed the *Trackmania* game, and Iliad, who revolutionized the Internet in France with the *Freebox*, have relied in part on active, creative and innovative user communities to renew their offers and develop new products and services. By opening up their products and services to these communities (and providing support to maintain the interests of users in the long term), they have been able to attract the most creative gamers. Communities have been organized into complementary user groups, based on their activities and contributions to the products and services, and this has also been a significant factor in the development of co-creation activities.

Creation toolkits and configuration systems are integrated into Nadeo and Iliad products, and the virtual nature of the service (based partly on software development and internet technologies) has given these firms great flexibility in incorporating users' ideas. Nadeo and Iliad also shared part of their value creation with the user community by moderating their prices or launching free products. On the other hand, Lego, which markets a more "physical" product, had to develop a new offer: Lego Ideas. Originally, customers could use 3D software to design any model, then have it made and delivered to their home. However, to meet the excessive manufacturing costs of specific products, Lego now only manufacture models under strict conditions of feasibility and economic profitability (i.e. models that are popular with users and that receive more than 10,000 votes in the website ideas.lego.com). This dialogue between the firm's expertise and user communities appears to be one of the key conditions for co-creation initiatives.

In the following sections, we present in more detail the practices observed in Nadeo's *Trackmania* communities and Iliad's *Freebox*, which have influenced the development of active, creative and innovative user communities.

I. A Game Based on an Active, Creative and Innovative Community of Gamers

In 2002, Florent Castelnerac, the CEO of Nadeo, a small video game design studio, asked his technical manager to develop a game concept that combined both track construction and driving cars. The idea was to enable the user to rediscover the sensation of being a child playing with small cars (but in a virtual environment). It is unlikely that Florent Castelnerac anticipated the consequences of this simple idea, which led to great economic success for the company following the involvement of a very active community of gamers.

At that time, Nadeo was a small firm of about ten people, mainly engineers and developers, who had already experienced great success with the *Virtual Skipper* game, a multiplayer sailing simulator that was very popular with sailing enthusiasts. Nadeo is a spin-off from the animation studio Duran, which stopped producing video games to focus on animated film production. The "video game" team at Nadeo worked with Duran's CEO to set up an ad hoc structure that could initially benefit from Duran's technology, premises and administrative framework.

In 2003, the video game industry in France was in crisis. An increase in the graphic power of consoles and graphics cards with each new generation caused a technological race forwards that was difficult to sustain for small development studios. The doubling of graphic quality and the volume of content in games required these studios to significantly increase their teams with each new production. This mode of development was only sustainable if the games had a high level of economic profitability and the studio had a solid financial base. Moreover, market access was most often only possible with the help of a publisher, in charge of game distribution, who was in a position of strength and took most of the margin.

Nadeo approached this market in reverse by deciding to turn away from graphic realism, focusing instead on the pleasure of the game and using a very refined graphic style. Nadeo's position as a firm supported by Duran also allowed it to be a producer and, therefore, to make decisions autonomously. Nadeo did not use a publisher, but financed the design and development of their games themselves and used distributors on a market-by-market basis. Focus, Nadeo's distributor in France, also contributed to the marketing costs of launching the games.

In 2003, with the help of Focus, Nadeo launched *Trackmania*, a simple and exciting car-racing game. Trackmania does not offer the complexity found in other games that try to get closer to reality. Instead, it provides a game that focuses on the pleasure of driving and features tools that allow gamers to create circuits and cars, and to organize multiplayer races online. When it was released, the game attracted attention with its simple concept, its racing-circuit creation feature and its attractive price (less than €30). It sold several thousand copies, which prompted Nadeo to develop a whole suite of games. In 2006, sales of *Trackmania* exceeded 500,000 copies across all versions. *Trackmania* then established itself internationally in a highly competitive market against "blockbusters" produced by world-class publishers. In 2007, *Trackmania* was the only motor-racing game present in the Video Game World Cup competitions and at the end of 2009, Nadeo was acquired by Ubisoft at a good price — proof of its exemplary success. Since 2009, four new versions of *Trackmania* have been launched. These are available on the Nintendo DS, Nintendo Wii and Sony PlayStation platforms and have enjoyed continued commercial and critical success.

However, though Nadeo's basic game idea and strategy (adapted to the highly competitive video game market) were undoubtedly brilliant, the success of *Trackmania* also depended on one major ingredient: the development of a very active community of gamers — fans of the game, who created activity, content and innovations. In May 2007, almost 3,000,000 gamer accounts were opened, and 45,000 gamers were registered on various French and international *Trackmania* forums. Every day, gamers organized more than 10,000 races and created more than 150,000 tracks. In 2016, the community was still as active as ever and, with the number of participants stabilizing, the publisher's forums had more than 55,000 registered gamers.

The first version of *Trackmania* did not include competitions, circuit exchanges or the opportunity to create videos of races. Innovative gamers took over the game and created the missing tools, websites and devices for themselves. Benz (the creator of the first competition league), Tom (the creator of the first Trackmania website) and Starbuck (the creator of funclips — a video competition from the game) are emblematic examples of these very creative users, who developed and encouraged activities within and around the game.

Initially, the gamers had no links with Nadeo; it was through their own initiative that they launched different projects, thanks to the creative possibilities offered by the game's toolbox. As Benz explains, their role was crucial at the beginning: *What's funny is that it's still me who invents the rules of the Trackmania competitions. We started on a particular basis, and four years later, it's still the same.* The community thus created ideas for innovations that were gradually integrated into the different versions of Trackmania: in-game automatic management of graphic resources, exchanges of circuits and access to gamers' sites. All these elements promote participation, contribution and emulation, which helps to make the Trackmania experience more exciting and social. The community has therefore played a major role in the game's success by providing it with rich content and fostering continuous innovation.

How did this community develop? What has been its contribution to innovation? And what was Nadeo's role? Observation of the Trackmania community (and other communities in the software and telecommunications sector, including the *Freebox* community) has allowed us to identify basic mechanisms that seem to support the development of a dynamic and innovative user community that is engaged in the creative development of the product it supports.

II. Opening up Products and Services to Attract the Most Active and Creative Users

The Trackmania community developed as a result of the game being open to user contributions (Parmentier and Gandia, 2013; Parmentier and Mangematin, 2014). Openness consists of providing entry points into the innovation process so that the user can participate in the initial design of the product or in its modification (to adapt it to their needs). The user is thus considered to be the co-creator in this process of innovation. The aim is to introduce user input at all stages of innovation, both in the generation of ideas and in the intermediate stages of development. At the technical level, openness takes the form of open-source software, toolkits for users or bulletin boards.

In Trackmania, this openness was concretized, for example, by user tests of the first playable versions during the development of the game and by the availability of a toolbox that allowed gamers to create content and new activities. Trackmania provides both content creation tools (circuits, cars, videos, sites) and activity creation tools (the organization of network races, local forums, chat during the game). The community created more than 150,000 circuits in three years, launched dozens of competitions and produced tens of thousands of videos (the most popular of which have been viewed more than two million times). At Iliad, openness is less important and users had little involvement in the initial design process. Nevertheless, the *Freebox* settings allow users to configure specific services and they take advantage of this to network their machines, produce new multimedia configurations, edit TV sites and broadcast their videos on TV Perso.[1] Iliad built on this by retaining ideas for improving and evolving the *Freebox*, which were discussed with community leaders

[1]TV perso is a personal TV channel managed by the user and accessible only with the Freebox.

during regular meetings at the firm's headquarters. Openness also concerns the identity of the firm. In the case of Trackmania and the *Freebox*, community sites use part of the original brand name — TM for Trackmania and Free for the *Freebox*. Free has gone so far as to lend domain names it owns to community sites.

Ultimately, openness attracts the most creative users in the community and involves them in the innovation process. In turn, these very active users contribute greatly to the development of the community by sharing their creations and knowledge. Their motivation is essentially intrinsic in the sense that they enjoy creating and innovating, and they (primarily) seek social recognition. They develop a strong reputation in the community, their contributions to forums are frequent and their names are known by those who follow community debates and events. These users support community development by creating websites and organizing new activities. Starbuck, who has launched several video production contests in Trackmania, explains his motivation: *What interested me was that developers start thinking about those who play their game, and who are as capable as they are of making the game evolve and create things for the game ... and that gamers can take it over.*

1. *The development of multiple relationships with the community and among users*

Nadeo established many relationships with the emerging Trackmania community from the beginning. A forum dedicated to the game was quickly set up and the studio manager himself responded to the gamers' requests. This kind of relationship does not happen naturally — the firm must create the conditions for it, based on a nascent community. The development of multiple relationships promotes the growth of both a community focused on the firm's offer and communication channels between the community and the firm. The relationship with the community is supported by technical devices (websites, forums, creative tools) and leaders who are well-identified and recognized by users. This association is necessary to capture innovative users who will participate in the co-creation of the innovation and promote the development of the user community. Interactions provide opportunities for meetings between users, content for forums and the emergence of website projects and activities related to the product and service offer.

Clash, a Trackmania gamer, explains the role of relationships in the game: *In the beginning, there weren't many of us, so relationships developed rapidly, we formed a core community that has remained very active up until today, that's what attracts me to this game. I joined the game, I connected to the net, and straight away gamers said "hi," "lol," "GG," which means that you take an interest and get hooked.* Since 2003, little by little, websites dedicated to the game have appeared and developed. For example, another user, Car Park, designs 3D models of cars with a wide range of associated skins, so that users can customize their vehicles. The firm has also promoted the emergence of major websites for gamers by financing their hosting, providing technical support and maintaining direct links with the managers of the most popular sites in the community. In the *Freebox* community, the major websites are still managed directly by users; for example, Freenews began as a personal page providing technical data; it became successful, and then developed into a professional site and a news channel on *Freebox*. Following its launch, the Freenews webmaster met regularly with Iliad's managers, and Iliad also financially assisted the project and hosted its servers free of charge. In 2008, Freenews attracted more than 600,000 visitors per month, while the TV channel was watched by 10,000 people per day.

III. Active Facilitation to Support a Dynamic Community

A community, even a virtual one, is a place to live. It only attracts and retains users if it provides a continuous flow of information and activities. To develop links and interest users in the life of the community, it needs regular events that bring users together and encourage encounters. Firms must therefore promote the organization of these events, directly or indirectly. Initiatives often come from the community itself and the role of the firm, in these cases, is to recognize their validity and help users to develop them by offering privileged information, financial and material resources, connection software functions, etc.

Facilitation encourages users to connect regularly to community sites and maintain their interest in the group, as Tom, creator of a website for the Trackmania community, explains: *We tried to show what was happening in the community, the little events, the new things... As people become interested, we try to give them info ... after that we proposed circuits, and vehicles. We also organized the first competition.* With Trackmania, users can easily organize games by switching their machines to server mode. The list of active servers and the number of players on each server appear in the game, and gamers can organize competitions and contests, as well as places to share their creations. Every season, TM Ligues, a popular game championship, offers a new circuit in which thousands of gamers try to obtain the best possible score. Players gather in small teams to participate in competitions. They divide the tasks between creators, managers and competitors to manage their servers, create cars with their own logos on them and plan training. The forums are also very active and dozens of new discussions are started every day.

In the *Freebox* community, facilitation is provided by forums and information sites. They bring together Internet users who are interested in the *Freebox* (or experiencing technical problems), and who discuss their problems, the *Freebox* offer and Iliad's policies. Regular additions of innovative functions animate the forums and are seen as major events. Meetings organized by Iliad with community leaders also provide a source of facilitation; they feed the sites with news, information and discussions. These events also make it possible to involve the most active members in the organization and animate the community itself. A dynamic community makes the product and service more interesting, which promotes its diffusion beyond the first generation of users.

IV. Interdependent and Complementary User Groups that Provide Mutual Value to Each Other

An active and innovative community is often organized into interdependent and complementary user groups, which have well-defined roles and use specific tools. The *Trackmania* community is made up of four categories of users: consumers, creators, managers and competitors. Consumers are the occasional gamers who fill the servers 24 hours a day. For the game to be interesting, there must be a wide variety of races running at all times. In addition, diversity of content is essential to make races exciting, renew gamers' interest and give a strong identity to competitive teams. In *Trackmania*, creators and managers contribute to this diversity. Creators assemble circuits, customize cars and edit videos. As Sam the Pirate, a creator for the game, says: *I created almost all the cars for my team. I made between 60 and 70 2D and 3D visuals and each of the cars is wearing our colours.* Managers organize international races and competitions. They have a key role because they are the ones who set up multiplayer races with their computers. For each race, they choose the circuits and lead the debates. They also manage teams of competitors by selecting gamers, organizing training and planning participation in competitions.

Competitors are regular gamers who participate in competitions organized by managers; they have only one goal: to win. Their presence is essential because without them the competitions would not be as successful, and the races would not be as intense. Competitions attract these gamers because they offer an opportunity for competition, in an organized way and with an indisputable ranking. In addition, competitions are a great means of communication because the "event" aspect ensures the promotion of the game to loyal and occasional gamers alike. These four groups are complementary in *Trackmania* and the community as a whole — they bring value to each other. For example, creators provide content to consumers and competitors to make races more interesting, and managers value their work by using it.

This principle of communities organizing themselves into complementary user groups is found in other content-based communities, such as the Wakfu[2] community in Ankama or the Wikipedia community. It also exists in the *Freebox* community, where leaders are clearly identified and invited to comment on the evolution of the *Freebox*. Administrators in various forums manage questions and answers about usability issues, and developers have adapted the open-source software, Freeplayer, to transform the *Freebox* into a real multimedia platform.

This method of organization leads to a categorization of users according to their skills and their contributions to the creation of the value of the innovative offer. It is facilitated by the provision of appropriate tools for each category of users, and forums for discussion and exchange. Creation tools allow each category of users to contribute according to their motivations and skills. Forums and download sites make these contributions accessible to all categories of users and allow them to be promoted by the community. It is then formed within the community of interdependent user groups in the sense that the presence and contributions of one group

---

[2]Wakfu is a massively multiplayer online role-playing video game (MMORPG) produced, developed and published by Ankama.

bring value to another group. Organization therefore promotes the growth of the community and the contribution of users to the design and development of products and services.

V. Fast and Direct Integration of User Creations

The rapid and direct integration of user creations and innovations supports the community's long-term goals. It promotes a sense of ownership of innovations, both within the community and the firm. Creation tools incorporate user contributions directly into the product and service without further development or the need to acquire specific technical knowledge. The use of application program interfaces (APIs) also allows advanced users to develop new features that are directly compatible with the product, thus avoiding conversion errors and costs.

Direct integration allows the community to evaluate user creations and this encourages the best innovators to get more involved, as explained by Scopius, *Trackmania's* circuit designer: *There are rewards that are attributed to the best circuits. It's true; I think I must be among one of the three or four top creators as far as the number of rewards is concerned. I know that my circuits are appreciated. So, I carry on, that motivates me to try to be just as good or even better.* The most interesting creations are downloaded more frequently; the others are forgotten. The most useful features are supported by many developers; the others are gradually abandoned due to a lack of support. Integration fosters community growth by providing valuable content to its members: the more valuable the community is to its members, the more it attracts new users. In *Trackmania*, gamers' creations are directly usable in network car racing and creators and managers contribute to this diversity. The game also offers features that reinforce the integration of community contributions: Manialinks and Maniazones. Manialinks displays gamers' websites and shared content directly in the game. Maniazones offers news, rankings and regional forums.

In 2005, Free launched its Freeplayer software, which is based on the open-source VLC software. Freeplayer allows the *Freebox* to be connected to a computer on the network to view images, music and videos. Subsequently, the community developed multiple versions of this software in order to transform the *Freebox* into a true multimedia platform. Today, Iliad takes ideas that are discussed during meetings with community leaders to develop them and integrate them into the *Freebox*.

VI. A Manifesto that Guides the Community's Path and Relationships with the Firm

A community needs to build an identity to justify its existence, to rationalize its relationships with the firm and to give meaning to its actions. The community acknowledges its existence, reflects on the meaning of its actions, and defines a purpose and objectives. The contributions of users to the design and production of the offer are justified by simple and understandable statements that can take the form of a manifesto. Newcomers to the community thus find the rules that justify the adoption of appropriate behaviour in the design of the offer. By affirming its values, the community gives an identity to the offer that is likely to attract other users. The creation of a manifesto is a process that promotes the dissemination of the offer to social groups interested in the values conveyed by the brand community. This manifesto can take the form of a user's charter or a debate on a forum. It can also be implied and naturally shared by members of the community.

For *Trackmania*, the manifesto, which is in this case implicit, is characterized by a common state of mind defended by community leaders, who themselves call it the "TM spirit". Gamers must give the best of themselves, whether they are competing or creating content, and they should share their creations and passions with other gamers while respecting the rules of good conduct. The TM spirit was not mandated by the game producer; it developed gradually from a core of highly active gamers who set an example by organizing competitions and developing a site for interaction and sharing. In parallel, gamers' involvement in creation and animation processes is well justified because all of them benefit from individual contributions. Moreover, gamers do not see the publication of *Trackmania* as a purely commercial operation; the perception is that the producer is there not to exploit gamers, but rather to enable them to have fun. Nadeo reinforces this impression by regularly publishing free additions to its commercialized games and freely disseminating several complete versions of *Trackmania.*

Within the *Freebox* community, Iliad is considered to be the most innovative access provider for Internet users, marketing the best offer, always at the same price (since launch), with no extra charges and no hidden costs. This, together with regular interventions by its CEO to defend Freenautes'[3] interests against shareholders, has strengthened the

[3] A Freenaute is a web user of the *Freebox*.

conviction among community members. As Benazech, who developed a multimedia player compatible with the *Freebox*, explains it, Iliad's position justifies his participation in information creation, technical support and debugging: *If I developed that, it was firstly because it was useful to me as well. And then after that, you don't develop for Free, but it's something that Free benefits from, it isn't a firm that has a bad image either ... they are very innovative, they like setting the cat amongst the pigeons.*

VII. A Sharing of Value Between the Community and the Firm

An active, creative and innovative community can have a strong impact on the firm's business model. The incorporation of creation tools into *Trackmania* is a way of acquiring additional resources that will bring value to the firm and the community, and initiate increasing returns.[4] The more gamers create content, the more gamers are attracted to the game. This requires the firm to reconsider its list of resources, establish links between the organization and the most active gamers, and change the way revenue is generated. The community thus becomes one of the firms' most valuable resources. The *Trackmania* game version, TMU, structures the community by geographical area and integrates it directly into the game using Manialinks and Maniazones. The producer has also recruited Benjisite, one of the community leaders who previously launched a blog and news radio channel on *Trackmania*, as a community manager to moderate the forums, distil information and highlight the most creative and active users. Other gamers, such as Tom, who developed some of the features of Manialinks and Maniazones, have joined the management team.

The firm's boundaries are therefore increasingly blurred and permeable. Producers do not hesitate to use a partially free model to attract and retain gamers, and then sell them an improved version of their game. Nadeo's business model is therefore based in part on valuing the "work" of users through the progressive construction of an increasingly successful offer, alternating between free (add-on and fully free versions) and

[4]The combination of economies of scale, learning effects and increasing utility related to the number of users results in the increased adoption of a product or service, and thus it spreads to a wider audience.

multiple sales of the same game (with an evolution of features). There is a tension between the need to charge for the service to acquire revenue and the need to involve the most active gamers in order to acquire additional resources.

VIII. A Virtuality that is Strengthened by Strong Links in the Physical World

With *Trackmania*, Nadeo has fostered the development of a strong community around a "virtual" product and service. The racing is indeed a simulation very far from real car racing and relationships between gamers are conducted via computers (although, with some encounters in real life). The game allows the creation of a virtual space that is very different from reality, but which offers a great deal of freedom to users and integrates the tools necessary to animate the community. *Trackmania* is an electronic sport (e-sport), and this shift towards "virtuality" has, paradoxically, allowed Nadeo to create strong and long-term relationships with its customers. The simulation of activities and the use of communication technologies make it possible to free oneself partially from the physical and temporal constraints of the "real" world. The very nature of the marketed product is then virtual, in the sense of "coming into being" through the succession of beta versions and marketed versions. However, *Trackmania* is also one of the games in the Video Game World Cup, where virtuality is transported into the physical world. Competitors are no longer avatars that you meet on the network; they are real people that others see playing the game. In addition, teams regularly organize local networking between their members in the physical world. This return to reality shows that there is a tension between the virtuality of an online community and its existence in reality. The community needs regular appointments in the physical world to ensure its real existence and continuity.

IX. The Innovation Process with a Community of Users

In these user communities, in relation to firms, innovation is a complex and recursive process that begins with the generation of user contributions ("contribution generation"). Next, ideas, content, websites and software prototypes are presented, discussed and enriched by the community and the firm ("contribution socialization"). Finally, these contributions are sometimes adopted by the firm and users ("contribution adoption"). For example, competitions were not included in the first version of *Trackmania*. They were set up by the gamers themselves and quickly attracted many fans. Following the gamers' comments and initiatives, these competitions have evolved and taken many forms, and Nadeo then incorporated features to facilitate the development and organization of competitions into later versions of the game.

Two of the practices presented mainly affect contribution generation: the active encouragement of a community ("animation") and the opening of the firm's innovation process to community contributions ("openness"). Contribution socialization is promoted through the development of many links between the community and the firm, and between community members themselves ("linking"). It also occurs via the organization of the community into complementary and interdependent user groups ("structuring"). Finally, collective reflection on community values and practices ("motivation"), and the rapid and direct integration of contributions into the innovative offer ("integration") encourage the adoption of user contributions by both the firm and the community.

Some practices are largely implemented by the firm, while others are supported by the user community. For example, openness is a firm-managed practice, since it is the firm's products and services that are open to user contributions. However, animation is more of a user-managed practice, although the firm can help users organize themselves and arrange events. Figure 1 illustrates the organization of practices that support both the development of a creative community and innovation with a community of users (figure adapted from Parmentier, 2015; see also Parmentier, 2016).

X. The Limits of Developing a Creative and Innovative User Community

The development of an active community of creators and innovators has enabled a small game producer to stand out against major international publishers. However, this method of managing innovation has several limitations.

While this type of management is possible for a small entrepreneurial firm, it is more difficult for a large, well-established organization to implement. The alternation between free and paid services creates income

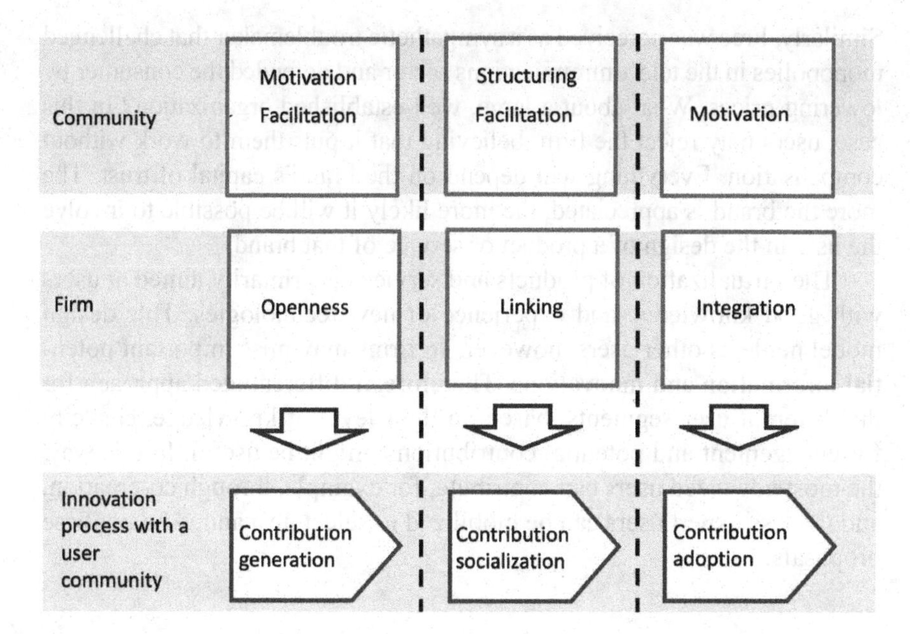

Figure 1: Organization and action of practices promoting innovation with a community of users.

uncertainties, and the larger an organization grows, the more it would aim to exclude such risk from its activity. Moreover, the opening of the organization to gamers is difficult to implement; it requires the appointment of a spokesperson and a partial questioning of the designers' expertise. It was possible in the *Trackmania* case because the spokesperson was initially the firm's director and later a community manager from within the gamer community.

The most active users tend to become more professional. Consequently, they must be remunerated for their creations or integrated into the firm. If this professionalization develops massively, the producer will not have the means to pay for creations. *Trackmania's* producer avoided this problem by creating a non-convertible currency that allowed creators to be paid virtually; however, a long-term consequence of this approach can be a monetarization of creations that takes precedence over the social value of creators.

The community quickly created content because *Trackmania* was a fun product and the producer was a small, "young" and "friendly" firm.

Similarly, Free was perceived as a sympathetic troublemaker that challenged monopolies in the telecommunications sector and defended the consumer by lowering prices. What about a large, well-established organization? In this case, users may reject the firm, believing that it puts them to work without compensation. Everything will depend on the brand's capital of trust. The more the brand is appreciated, the more likely it will be possible to involve the user in the design of a product or service of that brand.

The virtualization of products and services is primarily aimed at users with good knowledge and experience of new technologies. This design model neglects other users, however, so firms may miss important potential information and innovations. Therefore, a differentiated approach for the different user segments, based on their level of knowledge, capacity for engagement and potential contributions, might be useful. In this way, the most advanced users can contribute, for example, through co-creation, and the less expert users can be mobilized to select, test and validate these proposals.

XI. Conclusion

To build an innovation process involving a community of users, we have seen that a firm must finely manage its relations with this community by developing the practices of openness, networking, integration, animation, structuring and motivation. This innovation process then crosses the boundaries of the firm and the community, and benefits from the advantages of each type of organization. By developing this new skill in community relations management, firms create products and services that are more valuable to users, establish a more loyal user base, reduce their innovation costs and gain a competitive advantage that is difficult to imitate. With the rise of digital and communication technologies, a new competition is starting in which a firm must not only develop the best product in terms of features and prices, but also give it characteristics that allow users to take advantage and adapt it to their needs, thus giving meaning to their consumer acts.

References

Burger-Helmchen, T. and Cohendet, P. (2011). User communities and social software in the video game industry. *Long Range Planning*, *44*(5/6), 317–343.

Jeppesen, L. B. and Frederiksen, L. (2006), Why do users contribute to firm-hosted user communities? The case of computer-controlled music instruments. *Organisation Sciences*, *17*(1), 45–63.

Parmentier, G. (2015). How to innovate with a brand community. *Journal of Engineering and Technology Management*, *37*, 78–89.

Parmentier, G. (2016). How to innovate with a brand community. Youtube: https://youtu.be/F2gZhm6grB8.

Parmentier G. and Gandia, R. (2013). Managing sustainable innovation with a user community toolkit for innovation: The case of Trackmania. *Creativity and Innovation Management*, *22*(2), 195–208.

Parmentier, G. and Mangematin, V. (2014). Orchestrating innovation with user communities in the creative industries. *Technological Forecasting and Social Change*, (83), 40–53.

© 2021 World Scientific Publishing Company
https://doi.org/10.1142/9789811234286_0010

Chapter 10

Hacking Health: Building a Community of Innovation Through Events

Karl-Emanuel Dionne, Luc Sirois and Hugues Boulenger

Innovation lies at the intersection of different worlds, at the conjunction of foreign universes. The recombination of ideas and knowledge is the driving force of innovation (Ahuja and Lampert, 2001), be it within organizations or outside their boundaries. To produce new forms of innovation, this recombination process requires organizations to open up to new ideas and knowledge outside their normal scope while growing the capability to integrate them and make them fit their reality (Kaplan and Vakili, 2015). However, organizations' core resources and capabilities orient their activities and innovation inquiries (Coombs and Hull, 1998). This creates a paradigm, a set vision of the world, a sort of adherence to particular paths that hinders their ability to identify, assess and build on external ideas. It also hinders their ability to see and tap into rich sources of ideas and contributions, creative pockets that could otherwise bring them tremendous value (Carlile, 2002). Such creative pockets, bursting with valuable perspectives, initiatives and resources, thus remain untapped or even undiscovered, being too distributed, unstructured and always outside the confines of the organizations that could engage with them, bring them to life and bring them to market (Cohendet *et al.*, 2010).

The healthcare sector is a striking example of such dynamics, with strong institutions such as hospitals and pharmaceutical companies dominating their ecosystems while being almost impervious to sources of ideas and knowledge outside their traditional paths. This sector is however undergoing a profound transformation, driven by digital technologies and other massive drivers of change, that requires such institutions to increase their innovation capabilities Fast. The pressure is high. They must find new creative inputs to spark novel ideas and solutions to help them face their multiple challenges. Yet, their structure organized around rigid silos anchored by norms, regulations and standards, locks them in tradition, inertia and particular ways to innovate. It also makes it essentially impossible for individuals, communities and organizations outside their world to contribute to helping them, let alone to improve healthcare as a whole.

This is the challenge that Hacking Health has been tackling since 2012. Hacking Health (HH) is a fast-growing global movement that fosters innovation in healthcare through the construction of communities of innovation. Such communities bring together and connect a variety of contributors from multiple universes: (1) healthcare professionals, researchers and patients, (2) software developers, technologists, engineers, (3) UX, UI, graphic designers and artists, (4) entrepreneurs, investors, business people, healthcare and government administrators. Despite the divergent interests of each of these individuals, organizations and communities, Hacking Health brought them together in a large and dynamic community of innovation with a common purpose: Breaking down barriers and accelerating the pace of innovation in healthcare, enhancing health and well-being for all, and ultimately transforming healthcare systems around the world.

Now active in more than 40 cities on five continents, with the involvement of 180 passionate chapter leaders, nearly 600 volunteers, more than 7600 participants in events, which have triggered more than 1500 innovative projects, the Hacking Health movement grows a global impact through its action within local ecosystems via city-based chapters and leaders. Using strings of collaborative innovation events, they structure a *middleground*, a space of interaction between distributed innovators, filled with ideas and new technologies, and actors from the *upperground* who can contribute to developing and diffusing these ideas (Cohendet *et al*., 2010). Opening such a space produces a community of innovation.

To understand how Hacking Health achieves this *tour de force*, how they build and maintain a middleground and community, we will examine HH's history, methodology, organizational roles and particular approaches their leaders follow. We draw several examples and learnings from their

Montreal chapter, for practical reasons and given it was especially influential in the movements' history. Yet many HH chapters and leaders from around the world contributed to the edification of their unique methodology over the years and could all be equally cited and showcased in one way or another.

I. Why Hacking Health? The Drivers Behind the Creation of a Movement

The upperground-middleground-underground framework is a powerful analytical lens to help understand the striking needs that the Hacking Health movement addresses, and the tectonic plates and forces at work that drove its emergence. It shows that Hacking Health is an orchestrator in feeding the growth of the digital health market by organizing ideas from the underground and connecting them with upperground actors who could adopt or help commercialize them.

1. *Accelerating digitization of healthcare, and digital health market growth*

Digital and mobile technologies have been transforming lives, society and whole industries for decades. They changed the way consumers, citizens, professionals, communicate, learn, work, interact, relate to the world, stay informed and interact together and with organizations around them. Yet, the healthcare sector has been very slow to engage in such a revolution, falling farther and farther behind other service industries. Until the pressure became unbearable.

After years of stagnating, healthcare institutions were in dire need of new solutions to improve efficiency, quality of care, or simply to meet the ever-increasing expectations of digital citizens and sophisticated always connected patients. The gap resulting from their inertia, combined with the growing interest of consumers for health and wellness tools, has attracted increasing attention from technology creators, entrepreneurs and investors worldwide, and the healthtech market is now booming.

As demand grows, more and more opportunities arise to harness digital technologies for the modernization of healthcare service operations, treatment delivery and quality of care. And the penetration of technologies in healthcare markets, far from slowing down demand, creates new opportunities for technology creators. First waves of healthcare technology adoption become stepping stones for new ones to emerge and thrive.

For example, the digitization of patient records generated massive datasets, which in turn created gold mines for new AI-based medical discoveries and innovations that modern technologists are just barely starting to get their head around. The arrival of Cloud Computing opened the door for new modern electronic medical records to emerge, which in turn created new possibilities for exchanging data between healthcare professionals and patients that have yet to be fully exploited. The integration of health kits as a standard feature of Apple's iPhones and watches removed important technological barriers to entry to a slew of mobile apps creators, giving them the possibility to turn every phone and watch into a medical device. And so on and so forth. The North American health digital health market is now expected to reach 220 billion by 2025.[1] According to Startup Health, "2019 continued the strong upward trend in health innovation funding that we've been tracking for most of a decade. With $13.7B in total funding across 727 deals, 2019 was the second most-funded year ever, and we see the trend continuing."[2]

2. *The emergence of new healthcare models... and the increasing pressure to change*

A number of fundamental trends in healthcare models are reshaping how systems and individuals will be managing care in the years to come. One can only think of the shift to value-based healthcare, the establishment of accountable care in the US, or population health management or the ever-present and increasingly pertinent san graal of preventative health and health promotion. On the user side, patient engagement and health consumerism are emerging trends that are there to stay. The pressure grows for these new models of care to get established and achieve long-term impacts, which drives the need for continuous improvement at least, if not drastic changes and large-scale changes and innovation in this sector.

[1]According to Graphical Research, as published in April 2019.

[2]https://www.startuphealth.com/2019-q4-insights-report.

The time has come for the transformation of healthcare institutions, organizations and systems.

Despite these driving forces pushing healthcare towards digitization, this sector has yet to generate the outcomes we are all waiting for. Leading health firms still develop products that do not fit with their users, healthcare institutions are still closed to ideas external to their traditional innovation activities, limiting the ability to move towards digitization while promising ideas are still kept in the shadow of the upperground actors.

3. *The upperground*

The health industry upperground is constituted by *large corporations* and *healthcare institutions* that have the required resources, knowledge and legitimacy to develop innovative products. However, these actors, because of the rigidity of their processes and the stability of their business models, can't build on their capabilities to create ideas that connect with the needs of clinicians and patients.

Large corporations: Facing this plethora of opportunity, large technology corporations such as Qualcomm, Cisco Systems, Google and Amazon established pharmaceutical companies, and even the leading health IT enterprises such as Cerner, Allscripts Healthcare Solutions, eClinicalWorks are all grasping at straws to find new ways to create relevant new products and solutions that can perform on the health IT market. In addition, these corporations, despite their rhetoric, have a hard time truly understanding and addressing their users' needs in developing new technologies, which lead to the development of inadequate tools, ill adapted for clinicians and patients' real practices. Market dominance, rigid internal structures or just plain lack of market and technology insights systematically impede these corporations' ability to innovate, and confine them in their upperground, undisrupted by ideas that could spur projects with tremendous value.

Healthcare institutions: Healthcare institutions are by design focused on delivering care and services. In addition, academic healthcare institutions are also structured to drive the creation of scientific knowledge. In either case, these institutions are heavily structured organizations with clear sets of roles, responsibilities and interdependencies, governed by strict guidelines, rules and regulations meant for protecting patients' lives and well-being. The most progressive ones have clinical research and training agendas, which by nature put them at the forefront of medical innovation.

But in terms of management, while they have the resources and legitimacy to develop and capture value, they are by design in a gridlock. While their sophisticated structures and elaborate scope of activities across departments make them powerful operations, they are a typical broad-based upperground in terms of innovation and ability to change and adapt.

In particular, these institutions present a number of barriers that limit their adoption of digital health solutions. First, they typically rely on procedures to acquire generic technology solutions from established companies. However, buying "ready-to-use" or "off-the-shelf" technologies limits the adequacy of solutions to the user needs, context and circumstances of usage. Even worse, given the high level of validation required by most healthcare systems before purchasing any new technology, which result in extended processes of evaluation of already-developed technologies, they often end up acquiring solutions that are several generations behind. These validation processes require companies to present clear guidelines with predetermined objectives of scope, quality, time, cost and stakeholder satisfaction. Therefore, the healthcare system procurement processes and regulation severely impede or even prohibit true innovation, and magnify the inertia of institutions, so typical of upperground organizations. None of these approaches promote the exchange or use of knowledge between health professionals and technology corporations. Given the strict regulations, controls and highly specialized knowledge that are characteristic of the healthcare system, these institutions represent a challenging territory for external technological innovators and experts to contribute to its transformation. And internally, they typically don't have access to technological advances, knowledge, technological development know-how or processes, nor technological adoption, financing or valorization capabilities. They might have IT staff to support current software tools, but these institutions don't hire software developers to improve or develop new ones because it does not correspond to their core activities. They might have processes and staff to control operational conformance, but no process improvement or organizational change experts.

As a result, they continuously fall behind in terms of the generation and use of digital technologies and innovations that could at the minimum increase their efficiency, and enable them to tackle the transformations towards the new models of care. These shortcomings call for the exploration of new approaches to innovation in the health sector that are meant for feeding new ideas into the processes of these upperground actors to contribute to their transformation.

4. *Underground: The healthcare ecosystem is bursting with valuable creative pockets*

The digital health's underground is filled with distributed creative pockets that are bursting with insightful ideas. These creative pockets involve *healthcare professionals, physicians and clinicians,* but also *patients* as well as *technologists* and *startups*.

Healthcare professionals, physicians, clinicians: The keen observer can also see that, behind the curtains of rigid rules and procedures, the health community is filled with ideas and isolated improvement initiatives led by highly motivated healthcare professionals. Lifting the cover of rigid structures, one can uncover an underground of informal activities meant for improving the health services offered to the population. But given the organizational nature of hospitals and clinics, these activities take place outside formal roles and organizations. Prisoners of a heavy machine, these health professionals are isolated mavericks who try to make life better for their teams using light "hacks" they implement in their practices and sometimes those of close colleagues. From text messages to fancy Excel spreadsheets or sophisticated wall charts with sticky notes, these professionals try to respond to frontline challenges and needs to improve patient services or staff efficiency with imaginative ideas. Despite being built on low-tech means, their solutions are particularly valuable because they emerge from the challenges healthcare professionals face in their daily tasks or from close relations with their patients.

Patients: Patients more and more take an active role in their treatment. Facing challenges with the healthcare system, and with an unprecedented access to medical information thanks to the web, they are driven to take things in their own hands. "You spend 15 minutes with me per month, I spend 43800 minutes with myself during that time. I think I can contribute an insight or two", we hear again and again from patients in medical congresses these days. "Nothing about me without me" is the famous motto of participatory healthcare and patient inclusion advocates. Movements like "Patients Included" actively promote openness to patients' ideas, perspectives and contributions. International healthcare improvement expert on improving healthcare Don Berwick expresses it nicely, "we–patients, families, clinicians and the healthcare system as a whole — would all be much better off served if we professionals recalibrated our work such that we behaved with patients and families not as hosts in the healthcare system, but as guests in their lives" (Berwick, 2009).

Technologists, designers, entrepreneurs: Outside hospitals' boundaries, a growing number of entrepreneurs, software developers, engineers, UX designers, all masters of technology and modern solutions, aspire to play a role in improving health. They can invent a thousand and one technological tools and solutions, they can harness great people and rare talent in launching new agile teams, and they often do so with the dream of making a difference in their worlds. Yet they are systematically locked out of the healthcare sector because of its rigid structure. An entrepreneur interviewed, after participating in a Hacking Health event explained: "While digital chronic diseases solutions are just starting to emerge in institutions and pharmaceutical companies, I've had those ideas and solutions 15 years ago! We could be light years ahead. But there was no way to be heard..." Isolated inventors, engineers or software developers also try to do their part, but are confronted with an institutionalized structure and have no way to voice their ideas, fit in or contribute.

Startups: Startups, no matter how agile and tech savvy, no matter how brilliant and innovative, find it significantly challenging to adapt to the institutional health system. While quite efficient and creative, their traditional approaches could bring significant value to the healthcare sector, but their lack of understanding of the frontline issues, their lack of insights into clinicians and patients' realities, and their naïve conception of solutions make them struggle to bring about real transformations.

The barriers to entry into the healthcare system are so strong, actors outside the system, outside that upperground, cannot contribute to it. In order for these isolated creative pockets to be brought out into the open, for these underground players to be engaged and mobilized, and create ways for the upperground to hear and integrate their potential contributions, Hacking Health's co-founders and chapter leaders rapidly saw the essential need and opportunity to create and orchestrate a dance between these two disconnected spaces, and unleash new possibilities. Through its particular approach and methodology, it started connecting these upperground and underground people and resources to generate social and economic value, creating a middleground to break down barriers to innovation in healthcare.

5. *A middleground to break down barriers to innovation in healthcare*

A bottom-up experience to bridge the worlds of digital tech and health

Sparked by a highly motivated interdisciplinary group of doctors, clinicians, entrepreneurs, engaged citizens, engineers and software developers, the Hacking Health grassroots movement emerged in 2012 with the original intent of bridging the worlds of digital technologies and healthcare. While the issues to solve in the healthcare sector were numerous, the co-founders rapidly observed a very profound desire among technologists to contribute their skills for social good, and an optimistic open mind from upperground actors to explore new ways to "shake the house." This hunch raised a simple question: "when and where can people from these two completely separate worlds meet?" The group realized it had to create moments for these worlds to interact. They also developed the core belief that such moments would resonate with a deep human quest for purpose and value creation.

(i) Original hackathons

Instead of creating a new model format from scratch, the first Hacking Health leaders decided to adapt and modify a type of event already very popular in the tech communities: the hackathon. Hackathons are gatherings — structured as friendly competitions — of software developers and other technical specialists working together towards the goal of rapidly delivering a functional piece of software or hardware. The origin of the hackathon concept goes back several decades ago. Universities such as MIT and large technology firms first mobilized this approach to solve challenges related specifically to programming problems. Google and Facebook then popularized the approach with internal hackathons meant for quickly solving specific challenges in the development of their product features. Today, hackathons are organized by multiple entities for solving a variety of business and societal challenges, from cybersecurity to climate change issues, but that was not the case in the early days of Hacking Health who were among the first to use the model for a socio-economic purpose.

(ii) Hacking Health's hackathons

Hacking Health adapted the hackathon formula for the healthcare sector. While hackathons were originally technically oriented, unstructured gatherings of techies tinkering with new technologies, the Hacking Health version were designed to be co-creation forums aimed at triggering the emergence of new bottom-up initiatives to solve frontline problems identified by healthcare professionals and patients.

Moreover, Hacking Health's hackathons are essentially centred around the needs, ideas and frontline problems of healthcare professionals and patients. They are highly festive, high-profile events purposely aimed at winning the heart of both the upperground and underground communities and their notable leaders and VIPs. Their detailed format is described in "The Hacking Health Methodology" section that follows.

Hacking Health's first hackathons basically brought healthcare professionals and technology experts and programmers to work together towards this goal. The most recent versions have much more elaborate mixes of participants, which include patients, designers, entrepreneurs, researchers, healthcare administrators, business people and more such as graphical artists, musicians, politicians, decision makers, sponsors and collaborators of all sorts.

(iii) The original spark hitting a nerve

The first Hacking Health hackathon, set up by the organizers as an experiment, generated a response of a totally unexpected magnitude. Originally planned as a small local gathering in a Sauvé House room for 50 people, it attracted more than 300 healthcare professionals, programmers as well as business leaders, politicians and decision makers from all over the country. The organizers were stunned by this level of interest. They quickly had to find a solution to accommodate all these participants. Thanks to the emergency sponsorship of many upperground players who were open to creating new ways to support innovations, like electronic medical record vendor Nightingale®, the Business Development Bank of Canada (BDC), high profile VC firm Real Ventures and others, the organizers were able to rent a whole building to accommodate so many participants for a whole weekend.

Facing such level of enthusiasm, energy and motivation, organizers and participants collectively understood that the forum they had just created responded to a fundamental need in the sector. "We felt we touched a nerve," recalls cofounder Luc Sirois, "And with many global leaders wanting to galvanize their own communities in similar ways and bring Hacking Health to their cities, we saw a wonderful opportunity to all unite and create something bigger together."

(iv) Community leaders from around the world create a global movement

More and more community leaders from around the country and soon from cities around the world joined the original team and took upon themselves to become the builders of a Hacking Health movement that would span national boundaries. They organized Hacking Health Hackathons and related events in their own communities, perfecting the formula and methodology every time, rallying more and more upperground institutions and corporations to their inspiring quest, creating more and more social and economic value in their own cities, while remaining connected with the core group to share their experience and learnings with the help and cheerleading of a global core team. For all these leaders and their communities, Hacking Health was more than an event name, it became a global community of innovation they were proud to belong to, a rally cry to break down barriers to innovation in healthcare, to open the doors of the healthcare system upperground to the crowds of creative actors and technological experts of the underground.

Creating a "Middleground" through events

Without realizing it, and everywhere around the world, Hacking Health and its army of leaders slowly created Middlegrounds for innovation in healthcare. The middleground, this layer built on intermediate groups and communities, plays an essential interface role between the upperground and the underground, and produces a generative dynamic of creating ideas and bringing them to market.

Hacking Health builds a middleground by connecting the creative individuals of the underground with the institutions and corporations of the upperground. It builds a middleground with events and particular collaboration-oriented formats which open up temporary spaces where different professional communities can meet, interact and work together, building on a shared desire of making the world a better place through better healthcare. Indeed, Hacking Health gives members from different undergrounds, true pioneers in their fields, a voice and an opportunity to contribute new ideas, skills and practices. At the same time, it gives members of different uppergrounds new tools to spur creativity, insights and validation as well as a new source of ideas and innovative projects. Also, and this is a powerful contribution of Hacking Health events in the mobilization of upperground actors, it gives them a glimpse of what their future could look like if they decided to turn their own organizations into communities of innovation.

But what does it mean to connect underground and upperground actors? What role Hacking Health plays as a middleground actor to create an interface between these two separate worlds? And how is this role concretely played? In the following sections, we explore such questions by first presenting the Hacking Health methodology before highlighting some of the most powerful elements of its underlying approach.

II. The Hacking Health Methodology

In this section, we present the tactics at the center of Hacking Health's ability to create, feed and unite its communities of innovation to form a middleground through the organization of collaborative events. We focus on four types of events that were central in this process: the hackathons, the cafés, workshops, and design challenges. Table 1 recapitulates each event's description and purpose in the creation of a powerful middleground. In order to weave their own unique communities of innovation, Hacking Health chapters have developed, used, improved, augmented and combined these different types of events in a number of ways in their different regions of the world. But the core techniques remain surprisingly similar from one city to the other, from one country to the next.

1. HH hackathons: Trigger collaborations and action

Hackathons are Hacking Health's flagship events. The movement has organized over 150 hackathons across the globe and has therefore developed a real science in conducting these events to trigger collaborations and action.

The Mechanics of a Hacking Health Hackathon

From a very tactical point of view, Hacking Health hackathons are friendly competitions of typically around 48 hours, in which teams are assembled to develop technology-based solutions and compete for prizes awarded to the best projects. While the purpose of such forums is broader

Table 1: Recapitulating events' purpose in the creation of a powerful middleground.

Event types	Brief description	Middleground role
Cafés	Short monthly events, half conferences and half-networking events allowing participants to meet and learn from each other's fields (technology, healthcare, design, startups,…).	• Socializing peripheral actors and recurring community members. Recruiting new ones. • Provide a "low-risk" discovery experience. • Developing knowledge, a precondition to valuable creativity and innovation. • Exploring opportunities for innovation, ideas, problems to solve… • Engaging and inspiring community members.
Workshops and clinics	Hands-on events of various lengths and formats at the crossroads of lectures and "learning by doing."	Teach required skills for members' success: • Creativity, collaboration, communication (story telling, "pitch," etc.) skills upstream • Implementation and entrepreneurial skills downstream of events. • Underground skills ("the startup ways," creativity, etc.) among upperground members (healthcare professionals, administrators, …) • Empathy and design thinking among the underground members (engineers, etc.)
Hacking health hackathons	Lively co-creation forums structured as friendly competitions of multidisciplinary teams working together to rapidly deliver functional solutions to frontline problems identified by healthcare professionals and patients. Usually last 48 to 54 hours over a weekend.	• Create a deadline to maximize idea submission, a pressure cooker to get more done. • Trigger the emergence of new bottom-up innovation projects. Drive action-taking. Less talk, more action. • Bring experts from different fields together to maximize creativity and ability to innovate. • Energize members of the community of innovation. Appeal to and generate human emotions.

(*Continued*)

Table 1: (*Continued*).

Event types	Brief description	Middleground role
		• Create a buzz and build a reputation: showing the world what can be done when focusing on building simple solutions with the means at hand. • Create a low-risk space for structured experimentation. Lower the pressure of perfection. • Boost great projects and ideas, rapidly kill bad ones.
Design challenges	Similar to hackathons, design challenges span over multiple weeks, enabling the inclusion of trainings and workshops in the program.	In addition to the above, they provide time to • Validate solutions, prototypes along the way. • Develop a language and methods appealing to the upperground. • Reach out to, connect with and engage strategic upperground partners. Grow a network around projects. Trigger their interest for the project and the process. • Maximize the chances of adoption/implementation by upperground actors post-event.

than being just a competition, this notion is maintained on purpose to create a sense of urgency and motivation among participants, and to spur teams' collective engagement toward a shared goal.

Pitching in front of the crowd: Within the first hours of these events, healthcare professionals and patients who have ideas and challenges to be solved present them with a very short "pitch" in front of a large crowd of potential contributors — experts from software developers to entrepreneurs and designers — gathered for the occasion. Hundreds of such contributors come to Hacking Health hackathons with the hope of finding a good problem to solve and an interesting technical challenge to tackle. While some aspire to find the seed of a startup to create or join, the primary motivation for almost all such technology experts and entrepreneurs appears to be mainly the desire to put their talent and skills to good use for a meaningful cause. A variety of projects are presented by these healthcare professionals and patients — who become project leaders — depending on their disciplines and area of expertise, and on the size of the hackathon, which could lead to the presentation of 15 to 50 project ideas.

Team formation: After carefully listening to these pitches, crowd participants choose projects they wish to help and work with and gather around in a room organized as a matchmaking area for teams to be organically formed. Each participant selects a single team to be a part of, which comprises on an average 5 to 7 members who come from different disciplines. Participants select teams based on their interest regarding the project's intention, the coherence of their skills with the technical challenges underlying this intention and a fit regarding the distribution of roles and skills in the team. A natural selection happens here with many project ideas being left behind when not chosen by the crowd, leaving their leaders empty-ended. Many of them then combine projects or most often choose to join and contribute their skills to other teams with projects with similar orientations as theirs. As a point of reference, from over 50 pitched projects, about 35 are selected by the crowd.

The "hacking" stage: Over the following 48 hours, these newly formed groups work together to design and actually build prototypes of solutions to address the project leader's challenge or idea. During this phase, the team gets acquainted, learns more about the problem to be addressed and the skills of their new teammates. Then they work together on developing shared prototypes, contributing numerous ideas and techniques, sometimes called *hacks*, that are combined to form an overall solution.

The evaluation gate: At the end, each team is called upon to make a final public presentation and demonstration of their solution to a panel of expert judges, who select the best projects based on such criteria as value of the idea, quality of the solution, potential to succeed, quality of the presentation and of the team. These judges are selected by Hacking Health organizers to represent the different areas and actors involved in the upperground. Teams with the best overall projects win one of the many prizes. While in the early days prizes were mainly symbolic in nature, in the most recent years they could be of more important monetary value, although some believe monetary prizes could induce inappropriate pecuniary motivations or attract the wrong type of participants. Prizes representing support to help the teams actually get projects and ideas implemented in the upperground are unanimously seen as much more valuable when this support is real, in-depth, and led or fully embraced by healthcare institutions.

The value of Hacking Health Hackathons

Hacking Health hackathons contribute particularly well to orchestrating and shaping local communities of innovation because of (1) lowering risks and expectations by focusing on building simple solutions with the means at hand (2) their temporal structure that accelerate the development of ideas, (3) their appeal to human emotions and (4) their structured approach to organizing the experimentation of ideas.

HH Hackathons put things in motion by lowering the pressure of perfection: The notion of "hacking" here is very central to the success of the Hacking Health approach. "Hacking" means solving a problem with the means at hand, to divert the use of an object from its primary purpose so that it could resolve a problematic situation, whether in technology or in any areas of life. In this sense, Hacking Health Hackathon participants are not aiming at developing large-scale systems or perfect solutions, but rather small solutions that could make a difference in everyday lives of clinicians and patients alike. Hacking Health lowers risks and expectations by focusing on building small solutions and the imperative of taking action. "Perfect is the enemy of good" and in healthcare it is also the enemy of taking action. Proposing even just a hack to solve a situation, puts things in motion and has the potential to drive change often far beyond the original impulse.

HH Hackathons are high intensity pressure cookers that accelerate ideas and progress: The limited duration of the event is a key success factor. Time pressure bolsters creativity. Time boxing forces discipline and creates motivation. It creates a natural countdown to drive teams forward, make them work fast, take decisions fast and focus on developing a minimum viable product. The limited time encourages participants to quickly build on each other's strengths and to share openly in order to be more productive. Most importantly, as it will only last two to three days, such events facilitate the stakeholder's decisions to participate and to fully invest themselves. It creates a low-risk environment, provides rapid feedback and validation of ideas, by peers, the crowd and experts, and can eliminate months of working on poor ideas or poor premises.

Hacking Health Hackathons appeal to human emotions: Hacking Health signature events are purposely staged to create memorable experiences, emotional memories and a strong sense of belonging. They are true "Happenings" with a spectacular dimension, filled with festive music, fitness and relaxation activities, light shows, iconoclastic and mobilizing speeches. For participants and observers alike, for organizers and partners, the shared feeling of exhilaration continues long after the event. Several participants reported that this positive energy continued for several weeks and was even contagious to their colleagues when they returned to work. A hackathon calls on participants' inner desire to build a better world, and gives them a dose and energy to take action, be part of this community of innovation and yes, in many ways, become "innovation knights" as further explained in what follows.

Hacking Health Hackathons are structured experimentations: These innovation marathons, despite their festive spirits and the creative freedom provided to participants, are particularly well-structured experimentation environments. They build on the well-established design thinking methodologies as a framework of reference. Coaches ensure teams' progress through milestones over the weekend, sectorial or functional mentors deepen participants' perspectives on important matters at key moments of their pathways. At predetermined moments, swat teams of specialists provide constructive feedback to help teams improve their pitches, the visual design of their solutions and their final demonstrations. The result is more spectacular, the presentations hit the audience's imagination, and the process helps the team focus on what really matters for the development of their ideas in a context of digital health.

2. *HH cafés: Grow a community, connect universes and share knowledge*

Luc Sirois, cofounder of Hacking Health, recalls a critical encounter with Irene Pilipenko, who suggested the idea that Hacking Health could "federate a larger and more connected community with more regular events to attract members and keep everyone interested. More outreach to related communities would also be necessary." Irene naturally emerged as one of the first Hacking Health "Chapter Leaders" and the group started to organize a series of what they called Hacking Health Cafés.

HH Cafés are short events, half conferences and half-networking events that allow participants to meet and learn from each other. Held monthly, they maintain and grow members' interest in the Hacking Health community in general, and in particular in anticipation of the annual local hackathon. Each *HH Café* covers a specific topic related to healthcare issues or digital health innovation, such as innovation in geriatrics, or gaming in health, etc. A variety of speakers — researchers, entrepreneurs, doctors, community leaders, past hackathon participants, and others — present projects they work on, their progress, their view of the future of health, and lively discussions ensue. *HH Cafés* have a high "cost to benefit" ratio in terms of building communities of innovation. Compared to hackathons, they require a relatively limited investment of time and resources from organizers, but also from participants and collaborators, and yet they play an important role in the Hacking Health methodology.

To help federate a larger community and helping recruit more allies and participants, these *cafés* have three functions: to (1) socialize peripheral actors and recurring members, necessary for the initiation of collaborations, (2) develop a knowledge base among community members and explore issues with high potential for innovation and (3) recruit and engage the community members.

Socializing peripheral actors and recurring members

To maximize the value and impact of *HH cafés* on community building, a number of strategies are followed. First, they are typically co-organized with partners involved in relevant areas to establish or bolster connection and networking effects. They are organized according to local opportunities such as the demonstrated interest of these partners or other organizations that are strategically aligned with the Hacking Health's community of innovation, or based on particular needs of the chapter, such as recruiting members of a particular profile, for example. In addition, the *HH Cafés* are purposely held in locations relevant to healthcare, technological innovation or economic development — coworking spaces, startup incubators, hospitals, pharmacies, cool enterprise venues, etc. — creating in itself a form of discovery experience: the healthcare-related actors discover the tech, startup and business universe, and the technology-related ones get immersed in the clinical world.

The ease of organizing and participating in *HH Cafés* allows new participants and collaborators to explore collaboration with Hacking Health or their partners with no risk. It's quite easy for hospitals or foundations to host *HH cafés* and experience collaborating with the HH community, while having a chance to showcase their expertise and organization and potentially attract new collaborators.

Developing knowledge and exploring opportunities for innovation
Knowledge is an important precondition to creativity. HH Cafés are natural formats to educate community members on new technologies or health-related topics, and to trigger ideas for potential innovation. Speakers come from different areas of expertise and help bridge the knowledge gap between clinical, technological, political and research perspectives of the future. This sharing of knowledge creates a foundation on which community members build on to collaborate and generate new projects.

Recruiting and engaging community members
Speakers are selected for their expertise, but also for their forward-thinking views. This is done with the intention to inspire, motivate and engage participants. It can also provide energy and purpose to their personal mission and idea development processes. Moreover, as hackathons near, HH cafés become essential tactical means to raise awareness, recruit and prepare potential participants. This would be achieved by, for example, showcasing past hackathons' successful stories and highlighting the challenges to be featured in the next hackathon, by providing a showcase for upperground hackathon partners and for their challenges, and being very visible on social media.

3. *Improve skills and get in gear with workshops*

After a few years of organizing hackathons, HH community leaders wanted to stimulate the creativity and collaborative skills of participants upstream of the event, as well as their implementation and entrepreneurial skills downstream of the event. They were also looking for ways to help project leaders structure their innovative ideas in preparation of the pitch night, and boost project development capabilities among the newly formed hackathon teams. They launched *workshops and clinics,* a new event format at the crossroads of lectures and "learning by doing," to meet these objectives. Developing these tools was essential in connecting the underground actors who came from different disciplines to increase their ability to construct meaningful collaborations by reducing their distance in terms of the methods and language they relied on in their respective disciplinary practices.

The first clinics were built to introduce the "design thinking" and "user-centric solution design" concepts and methodology to health professionals, with very hands-on practice led by professional designers, real masters of the methodology. Personal definition, clinicians and patients journey mapping, friction points identification: All major concepts were taught and experimented to help healthcare professionals better define the problems to solve, to increase their ability to communicate them with future teammates not intimate with their context and environment. Future hands-on workshops introduced new technologies and were even extended to train community members on business or IP concepts.

Other pre-hackathon workshops rapidly emerged, such as "pitch" clinics, where seasoned entrepreneurs, investors or experimented hackathon participants would help project leaders shape up their stories and demos to be more compelling and convincing in the shortest amount of time possible. Healthcare professionals and researchers learned how to synthesize their ideas and present them in the most engaging ways to potential teammates through a variety of storytelling techniques. While such skills and practices sound quite straightforward in today's business world, they were and still are major novelties for clinical or research practitioners.

4. *A strong need for project implementation*

Despite the upstream effort made to improve the creative outcomes of the hackathons and the great momentum and number of positive contributions and support gathered during hackathons, the projects were struggling to become sustainable after the event. The initial structure of the community had not allowed an effective capture of the value created by its members, and while it generated more projects than ever to solve healthcare problems, too many of these projects failed to be implemented or come to full completion. While a fair amount of hackathon projects could not be maintained because of their lack of fit with the healthcare field orientations, others which deserved to be explored further encountered difficulties at the development and production stage, a critical step in achieving dissemination of the solutions developed. As Hacking Health aims at transforming the health system towards mobilizing more collaborative approaches for the production of digital innovations, it was essential to help sustain these projects, which could be effective vectors of change in the healthcare establishment.

After learning to federate different underground actors, Hacking Health had to develop methodologies for project groups to learn how to be recognized, legitimated and carried by upperground actors to effectively capture the value propositions developed within the community. For this reason, Hacking Health has deeply engaged from the outset with partners in the health sector, such as the CHUM and CHU Ste-Justine in Montreal, CHEO in Ottawa, Lyon Metropole and the University of Lyon Foundation in Lyon, CHU Besançon in Franche Comté in France and many others to not only be co-organizers of the events but so they would commit to pilot and eventually implement certain projects in their own institution.

Furthermore, to address this fundamental need to support projects post-hackathons, Hacking Health leaders of the most mature chapters launched full blown healthcare startup incubators, accelerators, living labs and other platforms dedicated to project implementation and value creation. Examples are the HH Accelerator and seed fund in Montreal, the iCare LAB in Lyon, the Vision Health Pioneers startup incubator in Berlin or most recently "La Couveuse" (the incubator) in Besançon. In other regions, the opposite happened: incubators and accelerators joined in the Hacking Health movement to bolster their community-building and idea-generation abilities. That was the case for Dutch Hacking Health, held in Nijmegen and multiple cities around the Netherlands, and organized by centres such as the ReShape Center at Radboud University Health Center, the ARK startup incubator in Valais, Switzerland, Innovacorp, Nova Scotia, Canada, TechTown in Detroit.

5. *HH design challenges: Capturing value to empower the community of innovation*

The experience of hackathons demonstrates the importance of prototyping ideas so they are quickly tested and become tangible means of communication for change makers. However, it also highlights the essential challenge of building a network of partners to maintain the momentum around nascent innovation projects. While hackathons helped spur creative ideas from the underground, there remains a need to connect underground ideas with upperground resources and validation opportunities. To address this, Hacking Health developed a longer version of hackathons, the Design Challenge. Started in Toronto and Vancouver, the model also starts with a pitch night and ends with a demo night, but spreads over several weeks. In Montreal, after a few years of partnering with Hacking Health for the organization of hackathons, Desjardins, a financial institution with a cooperative tradition, adopted the design challenge format that was later rebranded as the "Coopérathon." The Coopérathon was deployed with the help of Hacking Health in a number of other sectors, including education and environment. In addition to having similar functions than hackathons, design challenges also support project groups in (1) developing language and methods appealing to the upperground and (2) constructing a network around their projects.

Developing language and methods appealing to the upperground during which project teams work together intermittently and can attend a number of workshops and training sessions that are intentionally spread out over a temporal sequence that suits projects' development. In addition to learning how to collaborate across their disciplinary differences, team members had opportunities to develop the means for their projects to be legitimate in the eyes of upperground actors. What are the important evaluation criteria for venture capitalists or hospital institutions? How can a project's intellectual property be managed to promote healthy collaboration with corporations and research centres? These are all essential questions for projects to attract interest from upperground actors who have the required resources to bring these projects forward.

Constructing a network around projects: In addition to learning these tools, the design challenge format is meant to enable teams to find and build ties with strategic partners — institutions, foundations, research institutes, private companies, etc. — that could help validate their prototype and eventually pilot it, post-event.

The fundamental hypothesis of the design challenge format is to find support for projects' further development. By establishing strategic partnerships in parallel with the creation of prototypes, teams can adapt their development processes to align with strategic partners' interest, which becomes a means of enrolling these partners in the project. Projects therefore benefit from Hacking Health's community of innovation but also become active players in enhancing the community's ties with upperground actors. On the other side, by participating in the development of projects, upperground actors contribute resources in the form of their networks, financial means, knowledge and legitimacy that help projects go further and get eventually implemented.

6. *A Dynamic Series of Collaborative Events*

Hacking Health events, taken individually, certainly have an impact on their participants. However, it is by creating and building on an inter-event dynamic that Hacking Health has been able to federate a strong community of innovation. This dynamic series of events was iteratively developed from a variety of Hacking Health chapters' experiences and became increasingly effective. The organization of collaborative events structured in a sequential format allows for the construction of a community fabric that extends beyond the duration of these events.

This event dynamic includes cafés, workshops, hackathons and design challenges (see Table 1 for a brief recapitulation of these events). Cafés help build community ties by maintaining regular meetings between members of the community while stimulating the development of its knowledge base. Because of their monthly occurrence, they progressively gather an increasing variety of individuals and organizations to extend the reach of the community. The workshops focus on the development of collaborative and creative skills required to support the combination of this extended variety of contributors. These workshops prepare individuals to adapt their language and methods of creation to be more consistent with the principles of interdisciplinary collaboration and digital innovation in health. Hackathons then benefit from this upstream work that results in the creation of better outcomes in the events. The variety of perspectives and knowledge, and the ability to communicate across participants' differences feeds the strengths of the hackathon format. These events were essential in preparing the community so that it is ready to embrace the design challenge that requires a much more solid and broad community. Indeed, an important level of trust must exist with the local partners of Hacking Health to make them accept to get involved before and after the event. This investment requires, for example, structural and organizational preparations in collaboration with the partners without which it would be impossible to support projects coming from the Hacking Health community of innovation.

7. *Roles in Hacking Health's communities of innovation and events: Volunteers at the heart of the Middleground*

Hacking Health could not have created such a middleground without the thoughtful and systematic efforts of the highly motivated community leaders and volunteers who play a variety of essential roles in the mobilization of the uppergrounds and undergrounds. Table 2 briefly recapitulates the roles played by HH members in the creation of a powerful middleground. Hacking Health is structured as a global organization that relies on the effort of numerous chapters that mobilize the Hacking Health methodology to create and orchestrate their local communities of innovation. These chapters have the freedom of repurposing this methodology and adapting it to their local challenges as long as it remains compatible with Hacking Health's broader approach (which is presented in the following section) and mission. Within these two levels of orchestration, there are two broad types of roles: "organizational" roles and "event-related" roles.

Organizational roles

These organizational roles are related to the overall coordination and facilitation of the Hacking Health community of innovation in the pursuit and diffusion of its broader mission, and in the process of supporting the replication of the Hacking Health methodology and learning from the chapters' local experiments. These roles are meant to maintain coherence across chapters and through the different annual sequences of events. They are also meant to ensure the community of innovation evolves, integrates new practices and improves over time:

- *The guardians of the vision and values:* A decentralized organization, to be coherent and thriving, requires a clear alignment of each of its members on core values and visions. Members who are properly socialized to the organization's DNA will be able to somewhat independently perpetrate and perform the mission to move the organization forward. Having guardians of the vision and values of the organizations in that regard is foundational. At Hacking Health, this role is filled by active co-founders and "elders," respected community leaders and veteran organizers from around the country and the world who constantly demonstrated, over years of involvement, a commitment to achieving and protecting the mission and keeping the vision

Table 2: Recapitulating HH roles in the creation of a powerful middleground.

Role types	**Brief description**	**Middleground role**
Guardians of the vision and values	Elders, respected leaders and veterans ensure members live by the community values and act in ways compatible with the vision.	• Existence and constant reinforcement of a manifesto: a very important ingredient to a coherent and thriving middleground.
Global coordinator	Operational, finance and general secretarial responsibilities at the cross-chapter level. Is a key central actor for an otherwise decentralized, volunteer-based organization.	• Leadership, coordination and administration. • Ensuring a business model exists to sustain the middleground. • Seeding of new initiatives and cross-community efforts.
Chapter leaders	The coordinators of local chapters and the true chief local community orchestrators. True local and national leaders.	• Local community & chapter leadership. • Community, network & connection building. • Recruit, coordinate, motivate and inspire team members & organizers. • Engage the upperground. Build trust from their key actors.
Communication managers	Generators of content, visuals and information for all communication channels, from social media, to the web to newsletters.	• Promote and sustain a shared understanding of the manifesto. • Maximize event participation. • Promote success stories and make key members shine. • Maintain internal motivation. Inspire others to join the community.
Facilitators	Development leaders that connect, train and coach chapters and their leaders.	• Teach & get the manifesto engrained. • Keep a decentralized and distributed network of communities together. • Construct a healthy dialectic across communities and their leaders.

(*Continued*)

Table 2: (*Continued*)

Role types	Brief description	Middleground role
		• Extract and disseminate experience and learnings among leaders. • Codify practices and methodologies. • Train and coach leaders and chapters.
Outreach coordinators	Builders of relationships with all actors needed for the communities and events to thrive.	• Mobilizing the underground. • Ensure a wide array of actors engage and participate in events underground. • Tirelessly recruit new participants in the middleground.
Event leaders	Organizers of cafés, workshops, hackathons, etc. covering all dimensions from content, speakers to logistics.	• Make events happen, which are the space and moment where actors of the upperground and underground will meet, mingle and collaborate.
Volunteers managers	Recruiters and managers who, enroll, motivate and coordinate volunteers within the chapters.	• Provide the middleground with an ongoing ability to execute events. • Create and maintain a source of new organizers, coordinators and leaders.

alive. Both at the global and local levels, such guardians take on the responsibility to communicate and share the vision and manifesto of the global movement through different channels, in person at key local events and through their participation in key internal and external meetings. This maintains a sense of purpose within the movement, and triggers the interest of outside groups and organizations related to digital health innovation.

- *Global coordinator or director*: A global organization requires a level of leadership, coordination and administration that only full-time national coordinators or a global director could achieve. These roles help all chapter and community leaders and other actors to stay connected, synchronized and properly active in terms of playing their own role. The global coordinator will also play the lead role in the seeding, creating and managing of most cross-chapter initiatives, such as national competitions or ideation contests. He or she also oversees the management of most administrative, financial and global funding initiatives.

- *The Chapter Leaders*: The desire to bring innovation to healthcare, the sense of making a contribution, both social and from an economic development standpoint, is immensely local in nature. People want to solve healthcare challenges to help *their own* families and friends. People want to support *their own* lighthouse hospitals. People want to create startups for *their own* regional development. Local leaders and influencers by definition contribute to local networks and connections. They aspire to have an impact locally, for their own people. The global Hacking Health movement relies and builds on these local initiatives and motivations to form a meta community connecting these strong and active local communities so they have a global impact. Chapters and chapter leaders are therefore the fundamental atomic element of the global Hacking Health movement.

  Chapter Leaders are the coordinators of local chapters and truly act as the chief local community orchestrators. They assemble teams of local volunteers and organizers and together they become the anchors of Hacking Health in their cities or regions. With their intimate understanding of local realities, they are the ones defining the strategy to deliver the global vision in their own region. From a tactical point of view, Chapter leaders and their teams organize local hackathons, design challenges, cafés, workshops and training sessions. They take the best of the Hacking Health methodology and adapt them to their local specificities. They can also forge new ones and, with the help of the global chapter facilitator, and as part of the global leaders circle, feed them back to the global movement.
- *The Communication Manager*: Leaders of the movement often repeated: "Hacking Health does not physically exist. It is only people, ideas, values and communication. So we need to do all of these intensely." Communication plays a very important role in keeping the community together, in motivating its members and in attracting new ones. But for this to happen, finding the right words, using the right visuals are essential in promoting and sustaining a shared understanding of the Hacking Health spirit. Communication Managers generate content for social networks, as well as stories, reports, videos,

publications, newsletters. If done right, it makes the community of innovation shine, thrive and grow.

- *Chapter Facilitators or Global Development Leads*: To keep a decentralized movement and network of communities together, Chapter Facilitators codify and share the practices. By connecting with all local chapters, Chapter Facilitators communicate the Hacking Health way to new and evolving chapters and construct a healthy dialectic across chapters on their experiences and learnings. They connect local experiments with the evolving global guidelines, achievements and goals of the movement. Chapter Facilitators are gifted connectors. They are tireless cheerleaders of chapters' successes, wise coaches with a thousand and one tips and suggestions, and big-hearted developers of new friendships around the world. "People like Annie Lamontagne, who speaks nothing less than 7 languages, who is an international expert in corporate social responsibilities, has been the glue to Hacking Health's growth in the four corners of the world. Operating from the heart of Brazil for years, she tirelessly communicated with chapter leaders in all time zones, sparking their minds and their heart, and patiently collecting and sharing their know-how and experience worldwide," commented Sirois.

Event-related roles are played on a local basis by chapter teams. They focus on implementing all key steps to organize successful and high-impact events as well as to grow and promote the community. They include:

- *Outreach coordinators*: Working on building relationships with individuals needed for the Hacking Health communities and events to thrive.
- *Event leaders*: Each organizing one event on a rotation basis, so as to ensure a constant flow of events, cafés and workshops, throughout the year.
- *Volunteers managers*: Continuously enrolling, motivating and managing volunteers plays a critical role in providing the community with an ongoing ability to execute events, but also with a source of new organizers and leaders.
- *Communication managers*: Promoting events and success stories, building local followership and communities and helping their community members connect and shine. They are instrumental in building a reputation and in providing a loud and inspiring voice to their community.

These roles are essential in bringing the Hacking Health methodology to life and to federate agents of Middlegrounds. Without such a division of roles and responsibilities, crafted to address specific challenges and goals and adapted as needed to local contexts, it would be very challenging to rally such a diverse base of actors and stakeholders.

III. The Power of Hacking Health's Approach

Organizing a middleground can be achieved through a number of ways, and Hacking Health specializes in the use of events, most importantly strings of events and inter-event practices, for that purpose. So what are the key attributes of such events? What are the particular features that make Hacking Health events true middlegrounds to drive and structure communities of innovation?

1. *Bringing together diverse and complementary actors*

Connecting a diverse yet complementary set of contributors

Hacking Health sparks creativity by systematically bringing together actors that otherwise would never meet. By design, they bring together actors of divergent motivations but who can be aligned towards shared goals: (1) healthcare professionals who look for ways to improve quality of care but also to get their job done better and faster; (2) patients and their loved ones looking for cures, education, and ways to make their life easier; (3) engineers and programmers who want to grow their skills and put them to good use, build something meaningful. Some to find a purpose, some to gain financial returns on their knowledge investments; (4) designers who wish to push forward the art of user-friendly solutions in new fields and find new creative inspirations; (5) entrepreneurs who aspire to create successful new ventures, generate revenues and profits; (6) investors who look for robust opportunities to seed future profitable investments; (7) businesses that hope to find new ways to get their foot in the door to sell their products and services to hospitals; (8) administrators that aspire to find new ideas to make their institutions run more smoothly, reduce waste and costs and (9) even Hacking Health organizers themselves who, beyond their desire to bring more innovation in healthcare, often aspire to expand their professional networks and opportunities. All these contributors are essential in developing digital health innovations and transforming health systems.

Creating common grounds between divergent views

Connecting such various stakeholders is fundamental to enabling co-creation processes, but it's not enough. Creating common grounds is paramount for the creative process to really happen and for communities of innovation to form. This is essential as such stakeholders come with sometimes foreign, often completely divergent values, scheme of reference, languages, interests and even goals. Without common grounds, new interactions across these differences can lead to misunderstandings at best, or worse, to disagreements and conflicts. Nurses and tech entrepreneurs, for example, with their completely different frames of reference, would naturally have a hard time thinking alike. Without common grounds, they would never be capable of imagining the future the same way for any innovation, even if they built it together.

Hacking Health develops common grounds by creating moments in the city where the "normal" is suspended, where these contributors can let go of their institutional attachments and hierarchical chains to let loose and believe that everything is possible. Moments where stakeholders listen to and learn from each other, actively try to put themselves in the other stakeholders' shoes, agree to compromise and actively work together to spark the creation of new ideas. Moments in time that also aim to unite stakeholders around a common inspiring mission.

Uniting stakeholders around an inspiring mission

Hacking Health drives stakeholders' engagement and collaboration, realigns divergences and differences by giving a higher purpose to all through unifying rally cries. It has been expressed in many ways, "Breaking down barriers to innovation in healthcare," "Bringing innovation to healthcare," "Inventing tomorrow's health together." These expressions all amount to a higher calling: to improve health and well-being for all by pushing healthcare forward. This meta-mission that aligns the different stakeholders' orientation promotes a focus on the resources, means and knowledge each can contribute to the common mission rather than on their divergences. According to one of the sponsors, "Hacking Health has achieved what no one else has. Healthcare professionals and hospitals, for example, typically fear contacts and proximity with for-profit companies such as pharmaceuticals and financial institutions. But here, they become driven human beings working together."

Acting as a universal, credible, non-threatening mixer

Hacking Health's structure as a non-profit, volunteer-based entity that emerged from a bottom-up citizen initiative makes it a neutral zone that inspires trust and commitment, therefore contributing to the movement's ability to rally divergent stakeholders. According to Irene Pylypenko, this reinforced the notion of higher purpose, made the organization less threatening in the mix and became a crucial success factor over the years: "We are a neutral platform. We are not motivated by profit [...] and this is a clear advantage. Our goal is simply: health, innovation and collaboration. Hacking Health embodies an inspiring vision with global reach for every citizen as it aligns with the common good, unity against disease, cooperation, pluralism. It acts based on values of harmony and the human quest for meaning and purpose." Compared to firms acting as anchors in their ecosystems, the neutral characteristic of Hacking Health bolsters the community of innovation's focus on a shared meta-mission rather than on the strategic goals of specific members of an ecosystem.

Making everybody win

Hacking Health leaders are talented at recognizing and embracing the objectives, strategic orientations and roles of the community's different stakeholders and to adapt events to the needs of all. For example, when it comes to hospitals, Hacking Health emphasizes its ability to engage and give a voice to clinicians and staff to generate new improvement ideas that can lower costs and improve services, while for financial institutions and other large corporations, Hacking Health highlights the opportunity to drive social impact in a strategic sector while improving its brand. To interest investors, Hacking Health exposes the opportunity to discover future venture treasures in the local and emerging ecosystem of digital health that targets a market in full expansion. In addition to discursively addressing these different stakeholders' interests, Hacking Health's methodology, while offering creative freedom, also promotes the development of creative ideas and projects that align with the variety of interests of its community of innovation. For example, hackathon jury members are representatives of these different stakeholders and use evaluation criteria that intersect their different interests. This is why Hacking Health succeeds in bringing such a diversity of actors with divergent objectives: By making them converge.

2. *Organizing short events that drive the emergence of "innovation knights"*

Since the first Hacking Health Hackathon in 2012, more than 150 such events have been organized by HH chapters around the world. Each one brings together between 80 and 350 participants, not counting observers, associates from partner organizations, organizers, volunteers and other stakeholders. The highly decentralized structure of Hacking Health makes every event unique and particular. Yet, all of them share a number of key attributes and success factors. Each of them directly contribute to the creation and longevity of a true middleground.

Hacking Health events raise the interest of new partners and individuals to enroll them in the community of innovation. Inspired by the energy of the events and the human connections they spark and by the creative experiences they generate, many institutional partners from the upperground and participants from the underground will become involved in the community of innovation after the event by taking active roles in contributing to digital health innovation within Hacking Health or their own organizations. Indeed, their involvement is not only related to moving hackathon projects forward. Far from that. Their role in the community of innovation takes different forms depending on their objectives and skills, and the nature and degree of involvement changes over time. Most partners and collaborators of Hacking Health first participated in a hackathon, either as collaborators, organizers, participants or just observers. A 2015 hackathon participant, now in charge of innovation programs in a clinical academy, often repeats "There's life before the Hacking Health hackathon, and there's life after."

Institutions from the upperground

Partners in organizing events like cafés and hackathons usually become important institutional members of Hacking Health's community of innovation. In Canada, the BDC (Business Development Bank of Canada) was Hacking Health's first ally and later became their national sponsor and a connector to the startup and investment world. For them, getting involved in the community of innovation through Hacking Health was a unique opportunity to generate a new healthtech investment deal flow and even to trigger the creation of new ventures in the field. The Ste-Justine University Health Center (CHUSJ), on the other hand, chose to be the co-host with Hacking Health and HEC Montreal of what remains to date the world's largest health hackathon. At the onset, as explained by Dr. Brunet, then CEO of CHUSJ and now CEO of the Montreal University Health Center (CHUM), their goal with the hackathon was to bring out as many improvement and innovation ideas as possible from the front line of their institution, a feat they could hardly achieve without such a high-profile happening. Today the relationship has evolved and the CHUSJ and CHUM are more involved than ever in the local community of innovation. These institutions and their teams work closely with Hacking Health by supporting the validation and eventual implementation of projects stemming from hackathons, but also in the creation of a number of new ways to drive innovation in healthcare and make their community of innovation thrive. Dr. Brunet and his team are inspiring spokespersons of this community, and they help drive the adoption of the Hacking Health models and establish new collaborations and partnerships in communities in Canada and around the world.

Individuals from the underground

Hackathon participants often become the most active members of the community, whether they pursue their project or not. From the outset, individuals sign up for their first hackathon out of curiosity, if not almost naivety. They come with the vague intention to contribute to a good cause, and based on a general interest for healthcare, technology or innovation. However, by participating, they discover an exciting new world, and develop a strong connection with the values promoted by Hacking Health. The amount of progress made during these very intense events is so impressive, the level of energy and momentum injected into participants is so high and the newly formed relations so oddly strong that project leaders and teams walk away from them transformed, filled with energy and motivation to continue working on their project and make their ideas a reality. Most importantly, they walk away with the demonstrated notion that together with interdisciplinary teams they have the power to invent new solutions and make things happen, things that they would have thought impossible to achieve in such a period of time. They also come out of the hackathon with a material representation of their ideas, which help project leaders share it with colleagues and supervisors as they get back to their daily work routines. With that realization, many of them become "innovation knights," true agents of change in their own organizations and in the community of innovation.

Most often, past participants get involved in the community by attending other Hacking Health events, continue building or helping new relationships or even help organize or host some of them. Many will support new participants, informally or more formally as mentors, for example. Some will become ambassadors of Hacking Health or even more. Many will embrace the mission to transform the healthcare system and become real “innovation knights,” relentless change-makers and enablers in their own milieu. They then call for change within their organizations, roll up their sleeves and get involved in transformation projects, they actively work to break down barriers to innovation in their own environment. They become the Hacking Health community of innovation and embody its vision in their everyday reality.

IV. Conclusion

Benefiting from creative ideas that lie outside organizations’ scopes and boundaries is a great challenge for actors from the upperground while connecting with these actors is difficult for innovators outside their worlds. This chapter has shown how the Hacking Health movement built a lively community of innovation that became a middleground able to foster the distributed and disorganized creative pockets from the underground and connect them with the organizations that have the capabilities to push ideas forward and turn them into valuable outcomes. Hacking Health’s story is especially insightful regarding its event-based methodology to build middlegrounds, spaces of interactions for ideas from different disciplines to be combined, solidified and connected with the rigid institutions from the upperground. As Hacking Health leaders often say, the organization has given a voice to those who are kept in the shadow of official and stabilized models of innovation that are privileged in the healthcare sector. Hacking Health’s methodology helps such a generative dance to emerge between isolated undergrounds and a strict and closed upperground. Through its neutrality and versatility, Hacking Health has been able to integrate health institutions, attract the sympathy of several large organizations and create openness among multiple communities of professionals. Hacking Health’s unifying and meaningful mission has facilitated the creation of a common ground that aligned these various stakeholders’ actions towards transforming the healthcare system fabric.

To achieve this, Hacking Health develops dynamic and cohesive series of events that allow for the building of a community fabric that extends beyond the duration of these events. *HH cafés* allow to open institutional boundaries, to interest new participants and to integrate communities that are part of the underground. *Workshops* offer opportunities to strengthen shared practices that are essential to the development of the community of innovation. *Hackathons* facilitate ideation, project creation, team building and provide an occasion to span the boundaries between the underground and the upperground that rarely meet. In addition to the tight time constraints, it is the energy of the colorful moments of Hacking Health *hackathons* that propels this connectivity. The *Design Challenge* format as well as close partnerships with healthcare and other upperground institutions aim at solidifying a network of collaboration and partner organizations around the projects that are created during this event. By building on these different formats, Hacking Health has unravelled the true power of events. More than an event organizer, Hacking Health is a true orchestrator of a dynamic that lasts beyond its events and federates a community of innovation at the interface of a disorganized underground and a rigid upperground.

References

Ahuja, G. and Morris Lampert, C. (2001). Entrepreneurship in the large corporation: A longitudinal study of how established firms create breakthrough inventions. *Strategic Management Journal*, *22*(6–7), 521–543.

Carlile, P. R. (2002). A pragmatic view of knowledge and boundaries: Boundary objects in new product development. *Organization Science*, *13*(4), 442-455.

Cohendet, P., Grandadam, D. and Simon, L. (2010). The anatomy of the creative city. *Industry and Innovation*, *17*(1), 91–111.

Coombs, R. and Hull, R. (1998). 'Knowledge management practices' and path-dependency in innovation. *Research Policy*, *27*(3), 239–256.

Kaplan, S. and Vakili, K. (2015). The double-edged sword of recombination in breakthrough innovation. *Strategic Management Journal*, *36*(10), 1435–1457.

© 2021 World Scientific Publishing Company
https://doi.org/10.1142/9789811234286_0011

Chapter 11

Fertilizing the Indian Milieu of Ocular and Plastic Surgery in Delhi: The Role of a Global Agent

Karine Goglio-Primard, Odile de Saint Julien and Florence Crespin-Mazet

Developing the creativity of territories is a major challenge that increasingly questions policy makers and private organizations involved in the development of the economy and innovation at the level of a city, a country or a region. After having stressed the importance of attracting a creative class of workers (engineers, scientists, professors, architects, poets; Florida, 2002), the literature on creative territories has shown that local innovation is based on the existence and articulation between three creative layers. The upperground composed of formal companies, organizations and institutions; the underground composed of individuals and communities informally involved in various types of creative activities (artistic or scientific) and the middleground. Positioned as the central layer of creative territories, the middleground plays the pivotal role of fostering links and the transfer of ideas between upperground and underground layers. As developed in Chapter 1, it consists of various physical spaces (cafés, fab labs, co-design workshops) or cognitive places (platforms) favouring the construction of new ideas and their diffusion through various events and projects.

Our objective is to explore how an external actor stimulates territorial innovation through the development of its middleground layer. Based on an original case study of an open technology project in the field of ocular surgery, this chapter unveils how a South-African organization (Afrikaner) contributes to fertilizing the milieu of plastic and ocular and plastic surgery in Delhi (India) through a well-structured process based on three steps. Firstly, the identification of a fertile milieu for technology transfer; secondly, the structuration of its middleground through the creation of an incubator; and thirdly, the professionalization of both middleground and underground layers to enhance territorial connectivity.

After having introduced Afrikaner and its open technology strategy, we analyze its original processes and principles to help fertilize this milieu. We then review the key success factors that can be deducted from this rich case study and their implications for policy makers.

I. Afrikaner's Approach to Territorial Development

1. *Afrikaner and its strategy of open technology*

Afrikaner is a global company of South African origin, founded some 30 years ago by scientists to develop research in the fields of medicine and medical biotechnology. Graduates of major international universities, the founding members of Afrikaner have chosen to settle their firm near what is now known as the Silicon Africa valley because of the attractiveness of this territory in the life sciences. Due to its dynamism and innovation capacity, South Africa occupies an important position on the global scientific scene: pioneers in heart transplantation (1967), South African's medical communities and organizations actively participate in research against various diseases such as HIV/AIDS and malaria. As an active member of several of these communities, Afrikaner has made a significant contribution to scientific advances in life sciences and benefits from both a large network and undeniable recognition at the international level.

Afrikaner has two strategic activities. Its main activity is fundamental and applied research with the commercialization in the form of patents or licenses, of technological systems and scientific processes. Its related non-profit activity is the transfer of saturated technology to developing countries through Open Technology (OT) projects. In this activity,

Afrikaner mostly aims at creating a pool of local medical skills for humanitarian purposes while ensuring that its technology is fully exploited throughout the world (building its international legitimacy).

Through its historical activity, Afrikaner has strongly contributed to the development and revitalization of the South-African territory in the fields of medicine and medical biotechnologies. This experience enabled Afrikaner's managers to identify key levers supporting territorial development and to design a unique approach for enhancing the entrepreneurial and creative spirit of the developing territories benefitting from their OT projects. This approach considers entrepreneurship, in all its forms, as key to territorial development.

Step 1: Identifying a fertile milieu to leverage technology transfer

At the beginning of the year 2010, the Board members of Afrikaner decided to transfer laser technology exploited to design automated surgical processes based on artificial intelligence through an Open Technology (OT) project. In this humanitarian project, laser technology was mostly applied to ocular and plastic surgery (such as cataract therapy) — two medical specialties exploiting laser technology. Once this decision was made, two Board members — Robert and Sophie — were selected to conduct this project. The first step consisted in identifying a target country and a local milieu representing a high potential for fertilization thanks to the OT process.

Concerning the choice of the country, Afrikaner's took three criteria into account that ground the core of its manifesto: the country's developing nature, its strong medical need for the technology concerned and its capacity to enforce ethical rules (rules in force in the field + exploitation restricted to medical use). Based on the expert advice of several scientific colleagues from their international life sciences community, Robert and Sophie came up with a short-list of three countries (India, Turkey and Indonesia) having a need in ocular and plastic surgery. Thanks to their local presence and experience of Afrikaner in India for more than 20 years (business activity), they finally narrowed down their choice to India.

The choice of a target milieu for the OT transfer followed a second set of criteria related to the characteristics of the local territory in the domain concerned. The notion of milieu corresponds to a socio-spatial entity, geographically bound, in which various business and non-business actors play a role in a given activity (Cova, Mazet & Salle, 1994). In this OT project, Afrikaner was targeting a geographical territory holding a sufficient number of actors having the capacity to support or develop entrepreneurial projects in the field of ocular and plastic surgery.

This was far from guaranteed in a country where entrepreneurship is not valued and receives little support from policy makers. Sophie and Robert thus spent time analyzing the characteristics of the milieu of Delhi in the field of ocular and plastic surgery. The objective was not only to ensure that it qualified for their OT project, but also to identify local partners that could relay and amplify their action towards local stakeholders and spread the project after their departure (sustainability goal).

At the upperground level: Afrikaner needed to secure the partnership or support of two types of organizations. Firstly, qualified and reputable researchers in the academic field of medical sciences with knowledge of and respect for the profession's codes of ethics and up-to-date scientific standards. Secondly, local development aid associations and business angels that could provide funds or material support to entrepreneurs. The two project managers of Afrikaner identified Delhi as a potential target milieu due to their existing links with the academic medical community. They therefore contacted two Professors from the AIIMS University of Delhi and researchers from the University of Sharda, who expressed their interest in this project and their willingness to support Afrikaner by opening their local academic network and actively contributing to the selection of student candidates and projects. The academics confirmed that ADP — an association belonging to the University — could provide logistical and material support to entrepreneurial projects. In parallel, Sophie also identified several local business angels in the field.

At the underground level: Sophie and Robert were looking for communities that could help contact potential entrepreneurs in the medical surgeon community and transmit good practices (time saving objective). Their academic network guided them to the SSC — Student Surgeon Club — to relay a call for candidates among senior surgeon students. This community federates about ten associations of surgeons (plastic, ocular and cardiac surgery specialists) who regularly gathered to share their knowledge and improve their practice in their respective fields. SSC could be considered as a space for exchange, debate and mutual support between students specialized in the same surgical field. The objective of this call for candidates was to control the existence of a sufficient potential of students willing to join the incubator and exploit the laser technology for medical purposes. As mentioned previously, entrepreneurship is poorly integrated in higher education in India and thus rarely constitutes a deliberate career choice. The survey confirmed the interest and feasibility of Afrikaner's OT project with about 300 projects received. Afrikaner's technological and managerial support was received as a unique opportunity to learn and develop links to a renowned international organization and to develop their career.

The middleground level: This layer seemed to constitute the "weak link" of the local milieu as testified by the lack of exchange between its upperground layer (hospitals and medical centres, health organizations, universities and research labs, major international firms involved in healthcare treatment, global scientific organizations, international fund providers and innovation hunters) and underground layer (medical students and researchers in life sciences) at the time. This weakness seemed to stem both from the scarcity of middleground structures and platforms (places and spaces) in the field of ocular and plastic surgery but also the lack of exchange mechanisms across layers (events, projects). But far from being a redhibitory barrier, this structural weakness reinforced the relevance of Afrikaner's intervention by justifying the need for structuration and professionalization.

Hence, despite a culture granting little value to entrepreneurship, the milieu of ocular and plastic surgery in Delhi was officially targeted by Afrikaner's Board Members for their OT project. Its potential was testified by the existence of a creative class of senior surgeon students willing to explore the use of laser technology (strong entrepreneurial potential) as well as the commitment of formal structures in the upperground able to support these new projects and/or exploit the talents of its holders. The humanitarian project was justified both by medical needs in ocular and plastic surgery and the milieu's structural weaknesses at the middleground level.

In the OT project, Robert was then assigned the responsibility of technological aspects (technology transfer, international approvals, technical specifications, etc.). Sophie was responsible for the managerial organization of this transfer. This consisted of creating an incubator, supporting the creation of startups, qualifying business opportunities, financing and managing this incubator up until the creation of the last startup.

Step 2 — An incubator to structure the middleground

Due to the deficit of middleground structures in the local milieu, the second step in Afrikaner's process consisted in structuring and revitalizing the middleground by creating an incubator with local stakeholders. Called LaserIndia[1], this incubator was based in the premises of the AIIMS University in Delhi.

Through the creation of this formal structure, Robert and Sophie had a place (office) and a space available to coordinate their support to student entrepreneurs, organize technology transfer and monitor compliance with their management rules. As Sophie pointed out: *the incubator is essential to structure activities, strengthen relationships and save time because we can't stay very long.* The incubator formed a hub hosting and centralizing local and global resources both at the material level (the laser technology, the funds, the furniture and equipment such as computers) and at the immaterial level (expertise in entrepreneurship and in life sciences innovation).

The incubator testified Afrikaner's commitment towards local stakeholders and centralized their shared efforts. It proved essential to ensure the cooperation, commitment and ownership of the OT project (success factors) of the local academic community. As Robert pointed out: *We had to involve academics and show that we trusted them and that we respected their competence and vision. We were immediately in a logic of sharing and collaboration. We passed on our rules, but they also added their own.* Therefore, local partners were actively involved in the design and execution of the OT project.

The SSC organized the call for entrepreneurial project towards its various associations of surgeons: it received 300 proposals. Out of these 300 projects, the academics from AIIMS and Sharda universities had to select the most promising ones. Based on this short list of 150 projects, a jury composed of representatives from Afrikaner (Sophie & Robert), academics (from AIIMS & Sharda Universities) and the SSC finally selected 30 projects to be incubated within LaserIndia. They commonly agreed on the following selection criteria: the motivation of student candidates for entrepreneurship; their entrepreneurial intentions as well as

[1]A fake name for purposes of confidentiality.

their ability to exploit the skills and resources available through the incubator and its partner institutions.

On their side, development aid associations were responsible for providing material resources to entrepreneurs: funds, equipment and the premises to locate the incubator (the place). Among them, ADP freely provided two of its local assistants to the incubator. The role of these assistants consisted in helping Afrikaner's project managers communicate more efficiently with entrepreneurs and stakeholders and adapt their intervention to the local culture and customs. This decision proved very fruitful to avoid misunderstandings and create a climate of trust during the next stage of professionalization.

Step 3 — Professionalizing the lower and middle layers of the territory
Once the incubator was created and the projects selected, the next step consisted in professionalizing local actors. This professionalization concerned mostly entrepreneurs but also the two assistants — Shania and Dahia. The objective was to transfer them procedural knowledge that could sustain after the departure of Afrikaner's experts and could be applied to other projects.

Concerning entrepreneurs, the methodological support of Sophie and Robert focused both on the development of their technical knowledge in entrepreneurship (market research, strategy and business models, budgeting, elaboration of a business plan) as well as on their soft skills (capacity to design creative solutions and convince stakeholders). In this process, one of the key challenges was to stimulate the perseverance and the rigour required from project holders to be legitimized by their peers both locally and internationally (deep understanding of latest professional standards and codes of ethics).

To reach these professionalization goals, Afrikaner developed a set of operating rules that could be associated to a specific codebook. Among these rules were the regular and mandatory presence of entrepreneurs in the incubator and their active commitment to the business creation process: *Through the incubator, we could ensure that the founders of future startups were present, and that they met the milestones set even if it was difficult. It also allowed us to detect entrepreneurs facing difficulties and to rapidly help them* (Sophie).

However, enforcing these operating rules issued by a global actor also required that local entrepreneurs understood and accepted them. If the codebook did not align with local habits and customs (such as status of entrepreneurs; time perspective; business vs. family priorities), it could simply be rejected. This is why Afrikaner's support not only targeted student entrepreneurs but also the two local assistants. By selecting, recruiting and training Shania and Dania to the conduct of an incubator, and also by explaining its operating rules and negotiating their adaptation for increased resonance and sense-making with local actors, Afrikaner's action also contributed to professionalizing the middleground layer. Shania even went through a formal certification process in entrepreneurship. As real social, human and technical relays, these young "managers" kept the incubator alive by coordinating links with underground entrepreneurs (explanations of rules, creation of meaning) and other upperground structures, as well as by organizing the various events and projects at a technical, a logistical and a scientific level (see Table 1).

The professionalization process for entrepreneurs stretching over 14 months included various events and deliverables such as individual interviews and coaching sessions, go-no go meetings with academics and experts, preparation of pitches to business angels and private equity experts or venture capital fairs (see Table 1, step 3). These events punctuated by strict deadlines enabled to regularly structure and control the advancement of the various entrepreneurial projects and to rapidly identify sources of difficulties.

Apart from the scientific, managerial and social support provided by this multidisciplinary team (local staff and global experts), each entrepreneur incubated in LaserIndia also obtained access to university resources and skills from the Indian upperground (platforms, databases, software). This meant privileged links between the underground layer (senior students) and the upperground layer both at the local level (Delhi and India) and at the global level (Afrikaner and its global network in the medical community and in the private equity field).

It is thanks to this local and global support mixing rigour and flexibility that Afrikaner's OT initiative succeeded in stimulating the desire to learn and create for young entrepreneurs.

In terms of innovation, the results of this project can be summarized as follows:

— the creation of 30 innovative startups related to improvement of ocular and reconstructive surgical methods by laser in India,

Table 1: Main steps in the OT process.

Step	Actions
Step 1 Selection of a target milieu	**Birth and orientation of the OT project within Afrikaner**
	Decision to open the laser technology: creation of an OT project
	Choice of future uses & users of the technology (ocular surgery)
	Selection of India for technology transfer (among three target countries)
	Preliminary studies: Choice of Delhi as a potentially fertile milieu in ocular (and plastic surgery)
	Confirmation of the existence of a pool of senior surgeon students interested in entrepreneurship and willing to explore the use of laser technology for their professional practice
	Confirmation of the existence of partner organizations at the upperground level (academics, fund providers — business angels) willing to support the OT project.
	Official validation of the feasibility and interest of the project by Afrikaner Board Members
Step 2 Structuring the middleground	**Formalization of partnerships (contracts) with universities and academics**
	Various contractual clauses: respect for ethics, student-entrepreneur status, free use of development aid funds, access to various university resources and skills (platforms, databases, software), confidentiality.
	Call for projects issued by SSC: **300 projects** received
	1st selection of 150 projects by local academics (short-list)
	2nd selection of 30 projects by Afrikaner and local academics
	OT contract between local universities and Afrikaner (free local support from Afrikaner)
	Legal creation of the incubator by Afrikaner
	Interviews with selected project holders by Sophie & Robert: roles, objectives, rules
	Prior to final approval of entrepreneurial projects
	Students-incubator contracts
	Launch of the incubator — recruitment of 2 local assistants provided and financed by ADP.
	Technology transfer between Afrikaner and the incubator

(*Continued*)

Table 1: (*Continued*).

Step 3 Professionalizing the underground and middleground	**Development of entrepreneurial projects** through the incubator: — Support from Sophie & Robert (Afrikaner) but also Shania and Dahia (local assistants) to develop and control the consistency and assess the potential performance of projects — Entrepreneurs also benefit from the resources and competences of **local university platforms as well as aid fund providers** **Go-no go session:** short presentation of entrepreneurial projects in front of a jury composed of academics, the incubator members and local business angels: assessment based on the quality of the project (potential; technical and managerial feasibility; respect of norms; financial consistency) and the entrepreneur's profile (medical motivations; capacity to access local & international funding; dynamism) **95% projects obtain approval** for the next stage (startup creation) Afrikaner contacts their network of **international business angels** specializing in high-tech startups Support from the incubator (Sophie) to prepare venture capital fairs (private business angels) **Certification** of projects in each technological field concerned (obtained by Afrikaner) **Formal creation of startups:** Registration & payment of the various taxes of the new activity **End of the OT project:** South African stakeholders resign from the incubator (Sophie & Robert) when the last startup has been created

— the development of innovative projects related to the improvement of medical and scientific methods (patents) through partnerships with Indian medical centres,
— international partnerships for the co-development of US-funded mobile medical units in Mozambique and South African pharmaceutical companies.

Among the 30 startups, we can mention the success of two original social entrepreneurship projects consisting in the creation of mobile eye surgery "caravans" travelling to Indian villages to perform cataract operations free of charge. These 2 start-ups have been subsidized by the Indian state and sponsored by private firms for equipment maintenance.

The local upperground has thus been enriched through both bottom-up and top-down mechanisms:

— Bottom-up (exploration): Entrepreneurs fed the upperground through the creation of new eye surgery practices generating economic or social value creation (e.g. mobile practices).
— Top down (exploitation): Local and global upperground organizations have drawn on the pool of skilled entrepreneurs for recruitment or development purposes. At the local level, some entrepreneurs have been recruited by hospitals in India, after 2 to 3 years of operations of their startup. At the global level, both Afrikaner and other international talent hunters (e.g. major international consulting firms or business angels) have either bought out some of these startup firms or recruited their founders.

Finally, some global organizations engaged in co-development partnerships with some members of the underground (e.g. the Mozambique project).

After the departure of Afrikaner, this open technology project generated other positive externalities locally: the "effects" of the dynamization of this milieu continue today. For example, Shania, who initially worked with Afrikaner's team in the incubator, has created a web platform enabling to identify and connect all incubators in the field of life sciences in India. The platform (space) helps entrepreneurs obtain global certifications for their innovations.

All these initiatives show that the professionalization work carried out by Afrikaner and its local partners increased the international visibility of local actors from the underground and their capacity to develop cooperation on an equal foot print with actors from upperground organizations at both local and global levels. The success of the project has contributed to improve the local status of entrepreneurs and the desire to engage into new business creation in this field of life sciences.

II. Key Success Factors

Our case study illustrates that the creative potential of a local territory can be stimulated by an external actor through a structured process. It highlights several key success factors at each stage of these structuration and professionalization processes.

1. *A manifesto and a codebook informing a set of "glocal principles"*

The success of Afrikaner's action in stimulating the innovativeness of Delhi's territory in the field of ocular and plastic surgery seems to be closely linked to Afrikaner's capacity to transfer its practice in the field cumulated over 30 years from its South-African base. To manage this nonprofit activity, Afrikaner has formalized a set of well-defined principles and operating rules that guide its intervention and enable its project managers to delimit it in time.

These principles correspond to those of a manifesto and a codebook and inform different stages of the OT process.

The manifesto mostly informs the first stage of the process consisting of selecting a target country and milieu. It includes values and moral standards as well as principles for qualifying the milieu as a promising ground for fertilization (Table 2). These principles apply to any OT project carried by Afrikaner: they cannot be adapted or negotiated under any circumstances with local stakeholders. They inform Afrikaner's go-no go decision to transfer its technology in a given territory.

The enforcement of Afrikaner's manifesto is guaranteed by specific member(s) of Afrikaner's Board who regularly report to their fellow colleagues on this subject. In our case, Sophie and Robert shared the responsibility of the second set of principles (qualifying the milieu) while Robert

Table 2: The principles of Afrikaner's manifesto guiding its OT process.

1. Values and moral standards	2. Principles for qualifying the milieu as a promising ground for fertilization
Non-profit exploitation of the technology	—The existence of a sufficient number of qualified actors at the underground and upperground layers of the territory to relay and sustain Afrikaner's action locally —Upperground: Renowned academic organizations; business angels and fund providers —Underground: A critical mass of students involved in medical studies willing to engage in entrepreneurship; the existence of student communities (club, associations)
Respect of international codes of ethics and professional standards and in the medical field concerned by the technology	A need for territorial enrichment or "fertilization": — structural deficits in the middleground; lack of links and regular exchanges between the upperground and underground layers — the lack of entrepreneurial knowledge (need for professionalization)

ensured the respect of value and moral standards. Their legitimacy to play this role comes from their internationally recognized expertise in biotechnologies and Afrikaner's membership in the NHREC (National Health Research Ethics Council).

The codebook mostly informs the third stage of the process referred to as the professionalization stage. It guides the local exploitation of the technology (including its certification in different fields) and details various good practices in entrepreneurial project management. As depicted in Table 3, these rules include two sets of practices and principles of conduct considered as essential to be legitimized in the entrepreneurship milieu: they form a kind of "grammar of use" (Cowan, David & Dominique, 2000) for entrepreneurs and local managers that can be assimilated into a codebook.

In contrast to the manifesto, the content of the codebook is not "frozen" and has been negotiated and adapted with local stakeholders to increase its local resonance and sense-making and guarantee its local applicability. Consequently, local partners were also responsible for enforcing the codebook towards Afrikaner. The advantage of this

Table 3: The content of Afrikaner's codebook in Delhi's OT project.

Standards at play in the global entrepreneurship community	**Principles of conduct**
Standards for presenting projects (technology exploitation) — strategy — business models, — business plans, Guidelines for making a pitch and negotiating with business angels or private equities Guidelines for preparing the certification of the technology in their field	— Regular presence in the incubator — Attendance to face-to-face meetings with incubator's managers — Timely delivery of milestones — Regular reporting
Person responsible for their enforcement: Sophie (CEO Afrikaner — Professor and consultant in entrepreneurship & and consultant in strategy	Persons responsible for their enforcement: the two local managers (Shania and Dahia)

codebook is that it is applicable to other types of projects. It transmits both scientific standards and rules of good practice (behavioural) based on the accumulated experience of Afrikaner (a global player in the upperground) in OT project management.

Taking the analogy of business strategies, we could thus consider that Afrikaner adopts a "glocal" approach mixing global vision (manifesto) and integration of local culture and factors in implementing its OT project in a given territory (codebook).

III. The Creation of an Incubator

In our case, the creation of this incubator is the cornerstone of Afrikaner's approach and considered as a major condition for its local intervention. It activates the four mechanisms of a middleground namely — places, spaces, projects and events during the professionalization process. As a place, the incubator forms a physical base to activate the other three mechanisms. It therefore comes across as a key success factor for several reasons.

Firstly, as a legal structure, this incubator can formally engage in contractual agreement with the project global and local stakeholders:

Afrikaner (technology transfer), academic fund providers and aid associations (providing the office and the two assistants) and student entrepreneurs. Secondly, the incubator brings together in one place all the resources, skills and actors necessary for the development and success of the OT project: underground actors (student entrepreneurs), formal upperground organizations (academics, local researchers and global experts) and local ADP assistants. Thirdly, the incubator allows Afrikaner to implement one of its codebook rules: the regular mandatory presence of incubated entrepreneurs and their commitment to respect milestones. This rule enables to control the project advancement within a given timeframe and to identify potential need for support. Fourthly, the incubator forms the locus or hub that houses the technology and receives funds from public authorities and development aid associations. Finally, it forms a visible platform increasing the capacity and legitimacy of local actors to create global pipelines with international actors such as business angels, innovation hunters, global health organizations and certification agencies.

Hence, through the incubator, members from the three layers of the milieu could converge and connect. Worth noting is that this incubator is not meant to be a sustainable structure: it has actually been dismantled at the end of the OT process. In Afrikaner's approach, the creation of an incubator mostly aims at controlling the open technology transfer and the professionalization process of territorial actors (step 3). Through this ephemeral structure, Afrikaner's aims at transmitting its rules and methods through a practice-based, learning-by-doing approach and to enable local milieu actors to rapidly become autonomous.

1. *The professionalization process of local actors grounded in entrepreneurial practice*

The professionalization of the territory aimed at developing the entrepreneurial savvy of underground actors and at reinforcing the middleground's capacity to generate communication and open exchange mechanisms between its upper and lower layers. Through this process, Afrikaner's ultimate goal was to empower local actors to replicate and improve this knowledge to further nurture their territory in a sustainable perspective.

Concerning the underground, Afrikaner deliberately chose a learning-by-doing approach to leverage the creative power of its members. By working on technically-accessible projects (mature laser technology), student entrepreneurs could focus on acquiring the skills and attitudes required to create and manage innovative activities. It thus enabled them to gain confidence, recognition and external visibility in a relatively short-time frame, consistent with Afrikaner's agenda.

Concerning the middleground, Afrikaner also adopted a practice-based approach towards the two local assistant managers. By showing the example, explaining her practice and sharing her management tools and methods, Sophie greatly contributed to the professionalization of these two middleground actors. To increase their legitimacy towards the various local and global stakeholders, Sophie even pushed one of them to obtain a formal certification of her skills from a renowned external organization. All these actions proved fruitful from a long-term perspective as testified by the new middleground platform that she created after the end of the OT project.

Our case study thus enables to identify several key success factors concerning this professionalization process:

— An anchorage in low complexity projects at the technological and technical levels (such as those exploiting laser technology) to increase the efficiency of the process (focus on the main professionalization goal, time saving).
— A mixed management team composed of global and local actors to enforce the principles of the manifesto and codebook (see 1st set of key success factors), favour communication flow and the development of trust. This team adopts both formal hierarchical levers (formal control mechanisms by experts from Afrikaner and local universities) and community levers (enhancing transversal support across project holders).
— A focus on the acquisition of entrepreneurial skills (technical but also soft skills) considering entrepreneurship as a new way of innovating for the public good.
— A practice-based approach. The goal is to generate commitment, sense-making and true appropriation of Afrikaner's principles, managerial methods and skills by local actors.
— A steady pace of predefined activities based on several internal and external events pushing the entrepreneurs to commit to their project, and make regular progress: project deliverables, face-to-face meetings with management team, group presentations, venture capital fairs, go-no go sessions with academics, etc. This rhythm also enabled

the managing team to rapidly spot difficulties and provide adequate answers.

— The formal, contractual commitment of entrepreneurs to participate in these activities at the risk of being excluded from the incubator.
— A predefined deadline concerning the end of the local intervention of the global actor (in our case up until the creation of the 30th startup) to avoid generating dependency of local actors and pushing them to rapidly commit to the professionalization process.
— The access to the global actor's network in the life-sciences community to create global pipelines and increase the local milieu's renown and legitimacy.

IV. Implications for Policy Makers

This chapter opens new perspectives for public policy makers involved in the economic development of a territory and/or supporting developing countries. This chapter highlights the role that private actors can play in this process and the interest of encouraging mixed partnership initiatives involving various types of stakeholders (academics, private equity funds, associations, communities).

It suggests that the innovation and wealth of a local milieu can be stimulated by external actors with global experience and legitimacy in fertilizing other local milieus. Such external actors can bring their accumulated knowledge to the milieu through the transfer of both technological know-how and managerial insights in the development of innovative activities in a specific domain. This transfer relies on several key mechanisms, processes and behavioural guidelines framed by a manifesto and a codebook that increase its probability of success, leverage its efficiency (time, resources) and enforce ethics.

In our case, the creation of the incubator forms a cornerstone of this mixed public–private cooperation for local development. This incubator gathers the heterogeneous resources of a variety of public and private partners and fosters connectivity across the layers of the local milieu.

This work therefore suggests that public policy makers can encourage mixed forms of partnership with private actors to obtain support in developing specific milieus considered as holding a promising potential for development through fertilization. This means that policy makers should develop their capacity to carefully analyze the development potential of their local milieus to better target their territorial development initiatives.

This potential can be identified by the existence of a sufficient number of skilled individuals at the underground level and organizations able to exploit and grow these skills at the upperground level, in order to strengthen the links between them (need for structuration and professionalization).

It is precisely towards this middleground that both public and private partners should target their efforts as it constitutes the central layer supporting connectivity and cooperation between innovative individuals able to explore the use of a technology (application of their idiosyncratic skills) and formal organizations able to exploit them for business or economic purposes. In this process, they should first encourage the structuration of this middleground through the creation of various places and spaces able to host and concentrate the joint resources and knowledge of all actors involved at the local and global levels. Through our case, we have illustrated that incubators can form good examples of such hub structures.

Once this structural consolidation was achieved, their effort could focus on professionalizing local middleground and underground actors through training, coaching and support processes. As far as the middleground is concerned, the main objectives are, on the one hand, to stimulate the regular organization of events and projects and, on the other hand, to enforce the application of the codebook rules among creative individuals. It also aims at creating new connections with global upperground actors that can contribute to local milieu development. This is where the support of an external global actor can prove very fruitful.

Concerning the underground, the professionalization process consists in transferring good business practices (technical but also behavioural) to efficiently relate and communicate with upperground actors (local and global). The goal is to help creative individuals speak the language of organizations, adopt the implicit norms of upperground actors to be legitimized and therefore develop their ability to develop partnerships on a more equal footing. The case has illustrated that instilling entrepreneurial processes and culture can efficiently support this professionalization as it pushes innovation holders to better connect (creation of new pipelines) and understand the upperground layer. In our case, entrepreneurship is not the main goal but in fact a pretext or a useful step to local milieu development.

V. Conclusion

The case study has shown how a global player cooperates with local stakeholders to help leverage the creative potential of an existing milieu (i.e. the development of a specific activity in a geographically-bound territory) through a unique approach designed based on its own experience. This support system made it possible to densify and enrich the intermediate layer of the middleground, which proved to be a weak link in the territory through the creation of a new yet ephemeral structure.

Through its anchoring in a manifesto and a codebook, this original approach can be reproduced and continuously enriched to fertilize other territories. As exemplified by Afrikaner's experience both in India and other countries, it can be applied to various forms of innovative projects and, hence, contribute to continuously nurture territories from a sustainable perspective.

References

Cova, B., Mazet, F., and Salle, R. (1994). Milieu as a pertinent unit of analysis in project marketing. *International Business Review*, *5*(6), 647–664.

Cowan, R., David, P. A., and Dominique, F. (2000). The explicit Economics of Knowledge Codification and Tacitness. *Industrial and Corporate Change*, *9*(2), 211–253.

Florida, R. (2002). *The Rise of the Creative Class*, Basic Books.

Part 5

Communities of Innovation as Key Coordinating Modes to Develop Resilience and Creativity when Facing a Crisis

© 2021 World Scientific Publishing Company
https://doi.org/10.1142/9789811234286_0012

Chapter 12

Orchestrating External User Communities and Balancing Control and Autonomy in Fast Growing Community Contexts: Lego Group and Ankama

Émilie Ruiz, Romain Gandia and Sébastien Brion

The benefits of innovating with users are well established from both academic and managerial perspectives. Successful, well-known examples (e.g. LEGO, Procter & Gamble, Samsung) have induced many firms to pursue orchestrated efforts with user communities. Yet these efforts remain challenging: Unlike employees, users do not fall under the authority of the firm and are free to enter or leave user communities at will. An orchestration model, as proposed by Dhanaraj and Parkhe (2006), suggests a central actor might undertake a set of deliberate, purposeful actions to create and extract value from a network. But the vast promise of potential market growth and/or access to sticky and valuable knowledge sourced from potential customers (von Hippel, 1986) might tempt firms so much that they underestimate the coordination demands of user communities, especially as they grow.

Accordingly, innovation literature has stressed key issues related to the "creation" of a community, member participation and involvement, and the quality of users' contributions. Furthermore, both research and empirical evidence suggest some unaddressed tensions pertaining to the orchestration of user communities, because innovating with these communities demands regulatory and steering mechanisms that can stimulate, guide and capture the contributions of (many) users. In particular, orchestration might address two intertwined tensions. First, as a community develops, positive network externalities increase adoption rates but simultaneously make it more difficult for the host firm to impose a strong appropriability regime, which can jeopardize the value captured by the firm (West, 2003). Second, a variety–control trade-off arises for firm-hosted user communities (Jeppesen and Frederiksen, 2006). As the number of contributions from the community increases, varied innovation orientations may arise, such that multiple parties try to innovate simultaneously, leading to a loss of coherence in the system. New control mechanisms then may be required to restore coordination and alignment of the innovation trajectory with the host firm's strategic plan. Beyond emerging user communities, high growth communities appear strongly subject to such tensions.

By reviewing two successful real-world cases from creative industries (LEGO Group, a toy industry leader, and Ankama, a digital and transmedia creation firm), this chapter seeks to establish guidelines for how firms might orchestrate growing user communities for innovation. Our objective is to determine how firms that harness mature user communities can deal optimally with these tensions and thereby maintain the innovation activity over time. In particular, we find that formal mechanisms are essential to organize the creative process and community dialogue. If firms interact proactively with the community, they can guide users toward more value-creating behaviours. In turn, toolkits are key to support different stages (e.g. solicitation, idea generation, evaluation, selection) and create a circumscribed technological space for creative actions. The two cases also reveal ex ante and ex post formal mechanisms. The former organize and fix value creation conditions; the latter control and evaluate the results, to facilitate knowledge appropriation. Due to their complementarity, these mechanisms can manage and streamline the innovation process with communities. Finally, we show that informal ex post mechanisms support the evolution and growth of the community from a long-term innovation perspective. These mechanisms entail the deconcentration of governance in favour of communities, as well as defining structuring roles that empower users to leverage their own autonomy and self-control. Thus, the community learns to regulate itself, in accordance with the strategic orientation of the firm, which offers the further benefit of minimizing the coordination costs associated with community growth.

This chapter is organized as follows: In the next section, we describe key tensions identified by prior literature pertaining to user communities. We introduce the LEGO Group and Ankama cases, then analyze both firms' user communities, strategies and attempts to deal with the focal tensions and orchestrate their communities. Finally, we summarize lessons from those two successful cases, in the form of guidelines for addressing tensions and benefiting from user communities. These lessons pave the way to new perspectives on the challenges of coordinating with external communities for innovation.

I. Innovating with User Communities: Benefits, Limits and Tensions

1. *Benefits and limits of user community innovation*

Whereas early innovation management studies highlighted "communities of practice" (Lave and Wenger, 1991), current research acknowledges the various kinds of communities that can support innovation through interaction with firms. Sarazin *et al.* (2017) propose four categories of innovation communities: communities of practice, user communities, communities of interest, and epistemic communities. Among them, user communities are distinct in their high brand or product loyalty, passion and trust. Their members embody varied experiences, knowledge and uses of the product or services, yet they share common interests or goals in relation to that offering. When users gather, the firm can expect they will leverage their varied skills, knowledge and experiences to collaborate or compete in their effort to derive new ideas, promote the offering or solve a problem. The firm then benefits from its access to external sticky knowledge, costless promotions and appeal to potential customers, as well as its minimal cost outsourcing of some of the innovation process to consumers. For example, when Netflix challenged its community to improve its recommendation and personalization system, offering US$1 million to any team that could improve the predictive accuracy of its Cinematch recommendation software by 10%, approximately 51,000 participants from 186 countries enrolled. In turn, it improved its prediction system (and estimated its increased annual sales by nearly US$1 billion). Even in this competitive mode, the winning team that owns the solution shared it with the entire community.

To attract such users, firms need virtual tools, including a virtual platform, on which they might dedicate specific sections to new idea collection, problem solving or user interactions. Some of them also provide toolkits, designed to encourage and support users' creativity (von Hippel and Katz, 2002). For example, Procter & Gamble's "Connect + Develop" platform allows users to connect and participate in the company's innovation process, with two main sections. The "current needs" section reveals specific problems identified and pushed by the firm; the "submit your innovation" section instead allows community members to submit new original ideas, which Procter & Gamble potentially might use to identify future markets.

Despite these renowned and successful examples though, innovating with user communities is rarely easy. Prior literature identifies some prevalent challenges (e.g. Burger-Helmchen and Cohendet, 2011; Sarazin *et al.*, 2017), including firms' tendency to try to create instead of harnessing communities (Dahlander and Magnusson, 2008), the difficulties associated with maintaining or growing the community, and management challenges, including the high coordination costs needed. In line with empirical evidence, we propose that successful, mature firms that innovate with their user communities still must deal with two main tensions, too.

2. *Challenges for innovating with user communities: Balancing autonomy/control and orchestration/growth*

The sustainability of external communities depends on their level of openness (West, 2003). When a community is more open, the positive effects of its networks and its development spread more rapidly, so users are more interested in staying, to obtain the positive network externalities (Katz and Shapiro, 1985). Such network effects then can spur growth of open platforms, which also reduce users' fears of lock-in and switching costs, which should increase their competitive efforts, with promising outcomes for the firm's ability to capture value. Empirical studies of innovation openness indicate an inverted U-shape though (Boudreau, 2010; Laursen and Salter, 2005): Too much openness might lower the entry barrier so far that excessive competition enters the system, but insufficient openness leads to diminished knowledge variety or innovation. Even if the firm reaches the ideal middle of the inverted U-shaped curve, a large, autonomous community still may make it difficult for the firm to capture or absorb the many contributions offered, reducing the value it can obtain from communities (Shankar and Bayus, 2003). As this tension-oriented perspective indicates, firms that seek to benefit from communities confront an adoption–appropriability trade-off (West, 2003), and fast-growing communities may limit their capacity to capture the collective creation benefits.

Communities cannot be controlled like employees (Parker and Van Alstyne, 2018), because members do not submit to the authority of the company and are free to enter or leave as they wish. At best, the company can encourage but not compel actors to innovate. A few studies note the economic and contractual logic that explains the sustainability of external innovation communities, from users' point of view (Parker and Van Alstyne, 2018), but little research addresses governance methods by the firms that host communities. Researchers that have started to consider such questions (Burger-Helmchen and Cohendet, 2011) in turn call for further, in-depth insights into the appropriate level of control.

Furthermore, user communities are subject to specific dynamics, reflecting members' renewal and registration growth, active or passive profiles, and growing expertise (Piller *et al.*, 2010). Resolving the inherent tensions thus may require addressing the dynamic evolution of user communities and *ad hoc* orchestration. Boudreau (2010) shows that sharply reducing platform control can increase the innovation rate in specific contexts. But despite a general agreement about the need for control and behavioural regulation in communities, we know little about the appropriate level or types of control to exercise in different external community contexts. Such questions are especially notable because community evolution is rarely linear, so its developing growth and autonomy may feature varied temporality (i.e. slow or fast) and a complex trajectory. From this perspective, firms need detailed insights, based on empirical evidence, to be able to implement effective, sustainable community management. The rich cases we study in this chapter reveal some innovative ways to manage communities in the long term and deal with the two identified tensions.

II. The LEGO Group: Balancing Control and Creativity to Orchestrate a Growing User Community

The LEGO Group is a large, family-owned, Danish firm, founded in 1932 by Ole Kirk Kristiansen in Billund. Its famous product line, the

LEGO brick (which abbreviates *leg godt*, meaning play well in Danish), has twice been named "toy of the century." Originally, reflecting Kristiansen's former profession as a carpenter, the toys featured tubes ensconced in wood; in 1958, they started moving to the plastic version currently on the market. In the 1970–1990s, the LEGO Group underwent worldwide expansion, starting in Europe, then moving to the United States. The company currently employs more than 19,000 people, has a leading position in the toy industry, and earned 7.8 billion Danish kr. in net income in 2017. Yet even with these markers of success, the firm suffered a decade-long decline in the 1990s, such that by 2003, it was on the verge of bankruptcy. To address the crisis, the LEGO Group named a new CEO and sought to refocus on its main product. In addition, it adopted an open innovation strategy in an effort to exploit the innovation potential of its customers and users — a radical departure from its stance in 1998, when it sought to sue a Stanford student for broadcasting online how to hack the program for its robotic Lego Mindstorm® set.

The open innovation strategy instead led the LEGO Group to organize several different communities. For example, it quickly realized that many adults were joining, signalling that they still played with Lego bricks. Independent of the firm, many of them would meet, build and create together, sharing their passion for Lego bricks. Organic groups, calling themselves "Adult Fans of Lego" (AFOL), had popped up all around the world. The firm decided to accept and support the AFOL communities, but to manage them at least somewhat, it also created a "Lego Users Groups" (LUGs) network, which formalized the spontaneous AFOL group structures and provided an official gathering site for Lego users. In parallel, it created a high-status LEGO Ambassador Network, a community network that links the corporate LEGO Group with influential AFOLs. Along with a dedicated forum for the designated ambassadors, it provides a community locator, calendars and blogs for all AFOL, which in turn reveal activities and dialogues between the LEGO Group and AFOL. For the most creative and skilled AFOL, the firm created a "LEGO Certified Professionals" (LCP) status. These users have transformed their passion into an entrepreneurial, full- or part-time job. They are not LEGO employees, but they are officially recognized by the LEGO Group as trusted business partners and subcontractors. Finally, to support its innovation strategy, the firm created a crowdsourcing platform.

1. *From Lego Cuusoo to LEGO Ideas: Evolution of a growing community*

Formerly known as Lego Cuusoo, the crowdsourcing platform currently called LEGO Ideas allows users to submit new LEGO product suggestions. The main objective is to involve a wide range of users in the firm's innovation process, not just AFOLs. The LEGO Group employs a team of around 10 community managers. For our purposes, we focus on the LEGO Ideas platform, the firm's most important user community, to detail its evolution and the mechanisms that the LEGO Group has implemented to interact effectively with it.

Between 2008 and 2012, Lego Cuusoo was hosted by Lego Japan, in collaboration with Cuusoo, a Japanese crowdsourcing partner. If a contributor had an idea for a new Lego product, she or he could upload it to the open Lego Cuusoo platform. If it earned 1000 votes from supporters, registered within the Lego Cuusoo community, the Japanese branch office of the LEGO Group might decide to produce and sell it, as occurred for two sets: the Shinkai 6500, a Japanese submarine, in 2011 and Hayabusa, a Japanese space probe, in 2012. Both products achieved commercial success. As a beta test platform, Lego Cuusoo struggled to manage the community, which reached more than half a million members who submitted ideas and votes. Therefore, and noting the success of the platform, the Danish headquarters decided to launch a more robust crowdsourcing platform that also would extend beyond Japan. In 2014, Lego Cuusoo officially became LEGO Ideas. Users were invited to submit new product ideas, using a dedicated toolkit. Community members could vote for these ideas; if they attained 10,000 votes, the LEGO Group's board would decide whether to put the product idea into production. An idea commercialized as a set for sale earned the idea owner 1% of the royalties on those sales.

Beyond voting, a specific interaction forum section allowed users to chat with one another and firm-employed moderators. By July 2019, 27 sets initiated on the LEGO Ideas platform had been commercialized, including a Minecraft set, which earned 10,000 votes within two days (cf. the Shinkai 6500 set, which got 1,000 votes within 420 days, and the Hayabusa set, which reached 1,000 votes after 77 days). Then a more recent revision to the platform created three modes for interacting, differentiated by the members' own profiles (e.g. skills, involvement, motivation). Specifically, members of the community can

- Submit an idea (available since 2014, without any imposed theme), as described previously, which is the main mode used by experienced, skilled and highly motivated members. It targets "master-builders" and is not limited over time; it promises the potential of net income as a reward.
- Enter a contest, in competition with other community members, with some dedicated theme (e.g. "Re-creating a Magical Harry Potter™ Holiday Scene!"). Users may be rewarded with Lego sets. Thus, the firm encourages these community members to be as creative as they want, which also represents their feedback, pertaining to specific topics that the LEGO Group wants to address. This mode is open to everyone, regardless of skill level, and simply requires interest in the predetermined theme.
- Undertake activities, after being invited to join the community in developing creative skills through fun challenges and exercises, without reward. All members of the community can participate, beyond submitting ideas or voting, regardless of skill or involvement level. This mode is largely for beginners and users with limited time.

2. *Supportive mechanisms to deal with control and creativity*

The tripartite structure of the platform helps the LEGO Group involve more users more actively, as well as support their skill development, by issuing simple exercises and challenges that beginners can enjoy before they consider entering a contest or submitting an idea. This organization gives the firm a means to implement formal and informal mechanisms for orchestrating its growing user community:

- **Opening the toolkit to increase autonomy and creativity:** To control and harmonize the quality of users' submitted ideas, the LEGO Group previously required them to use a specific toolkit, Lego Digital Designer (LDD). They had to download LDD, learn how to use it, and spend time interacting with a software they probably had not used before, so the entry barriers likely were too high for many creative users. To reach more users and encourage more creativity, the LEGO Group now accepts any design toolkits; people can even use actual bricks to present their designs. This major change implies a de-structuration of the user community innovation process, yet the formal mechanism actually gives users more creative autonomy.

- ***Refining guidelines to control the process.*** The LEGO Group has always provided explicit process guidelines, yet despite their transparency, these guidelines tended to be poorly described, specifying only four broad steps for submitting an idea: (1) share your idea, (2) gather support, (3) LEGO review and (4) new LEGO product. A revised version instead specifies nine steps: (1) have a great idea; (2) read the guidelines (with a link to a "rules" internet page); (3) check for intellectual property licenses, because many submitted ideas refer to existing movies, series or videogames; (4) build it, with whatever tool; (5) submit, with the possibility to add more content; (6) engage with the community; (7) 10K Club (see the next paragraph); (8) approved for production, meaning that internal LEGO designers will help the user finalize the idea for approval by the firm; and (9) glory and recognition. As these guidelines indicate, the firm thus provides users with more advice, resources and support to help them submit viable ideas.
- ***Diversifying users' roles to expand the community.*** In addition to structuring the community and controlling the process, the LEGO Group grants users other means to get involved. The new activities section on the platform encourages varied user interactions, to help foster the essence of the LEGO Ideas platform, namely, the community. A prime example is the dedicated status, "10K Club Member." If a user submits an idea and attains 10,000 votes, this user becomes identified as a member of a very specific subgroup in the community. Advice provided by the LEGO Group suggests ways that motivated users can reach this status, including the explicit assertion that *becoming a LEGO Fan designer is incredibly challenging and requires a unique brick-built concept, solid planning, a boatload of determination, as well as a healthy amount of patience.* In terms of timing, the firm seeks to educate users that reaching a high status requires lots of time and steps (Figure 1). With these efforts, the firm not only encourages the submission of qualitative ideas but also addresses crowding effects (Piezunka and Dahlander, 2015), which arise when firms receive too many poor ideas from an unstructured crowdsourcing platform.

As these mechanisms indicate, as the LEGO Idea community grew (to around 1,000,000 members registered today), the firm increasingly

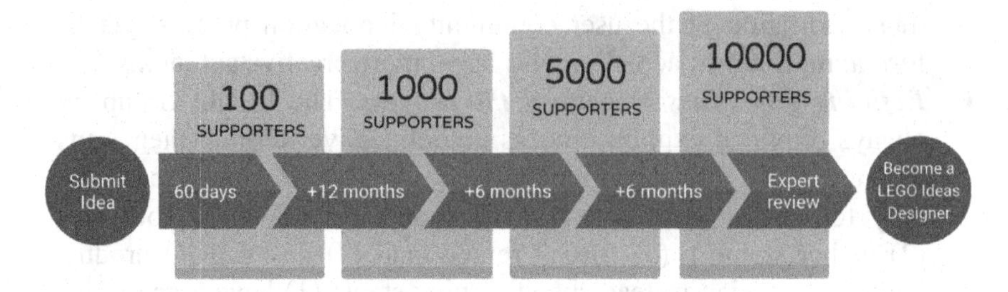

Figure 1: Becoming a LEGO Fan designer.

Source: https://ideas.lego.com/projects/create.

structured the submission process and community members' roles. In so doing, it could better control for quality and the volume of ideas, while also granting more flexibility and autonomy to users and encouraging their creative efforts and persistent involvement over time.

III. Ankama's Orchestration Model: Control and Empowerment of Community

Ankama is an independent digital creation group, working in the field of transmedia entertainment. Founded in 2001 as a Web agency (i.e. interactive communication activity), the company quickly diversified into the massively multiplayer online video game sector with its first success: *Dofus*. Launched in 2004, *Dofus* offers a very open, collaborative medieval fantasy universe, targeting 12–25-year-old players, in which the multiple future narrative evolutions encourage players to engage quickly. In less than a year, more than 450,000 players joined *Dofus* and formed the first community in Ankama. The company continued its diversification between 2005 and 2007, by developing a press and publishing division (Ankama Editions, for books, magazines, art and manga related to the Dofus universe), a products division (Ankama Products), and an events division (Ankama Convention). At the end of 2007, Ankama also introduced *Wakfu*, a new massively multiplayer game that features the Dofus universe, set 1000 years later (it remained in beta testing for four years before officially being released in 2012); the voluntary connection encouraged the Dofus community to join the new game, too. Furthermore, it is working on a new animation division (Ankama Animations, managed since 2009 by Ankama Studios in Japan) to create a televised series, connected to *Wakfu*. Its objective is to create a true transmedia universe, filled with multiple complementary consumer experiences, across three primary media: (1) the online game, exploring the present; (2) the animated television series, exploring the future; and (3) comics and manga that recount the characters' past. Through these narrative and temporal interconnections, Ankama expands its community, such as by attracting viewers from the television market. Between 2010 and 2012, the company initiated the international distribution of *Dofus* and *Wakfu* games, in Europe (e.g. Spain, England, Belgium, Germany) and South America (e.g. Brazil). In France, *Dofus* maintains 30 million accounts and 3.5 million subscribers — surpassing *World of Warcraft* as the most popular game in the country. *Wakfu*, which officially launched in 2012, is available on Xbox consoles and mobile devices. The television series also has achieved international success (particularly in the Asia-Pacific and on the Steam platform). In parallel, Ankama Music manages music from Dofus and Wakfu, and a media division is responsible for Web TV and video-on-demand distribution. The Ankama Convention has become a reference event in France, attracting 20,000 visitors over two days, who can discover and test new products, meet artists and compete in online game competitions. More international studios continue to be added (e.g. Ankama Canada in 2013, Ankama Singapore and Brazil in 2014, Ankama Asia in 2016), and animated films set in the worlds of *Dofus* and *Wakfu* have been released, along with a *Wakfu* series available for streaming on Netflix. Together, Dofus and Wakfu maintain more than 60 million accounts and 7 million active users (more than half of them international). Ankama generates more than 70% of its turnover from video games (€40 million on average), with a workforce of around 500 people in 15 companies around the world.

1. *Evolution of orchestration mechanisms due to growth in Ankama communities*

In ten years, Ankama emerged as a massive, independent digital creation group that runs a vast, transmedia product ecosystem to satisfy a community of several tens of millions of users. This atypical growth rate required dedicated community management efforts to deal with the rapid expansion (for Dofus, from 450,000 players in 2004 to 3 million in 2007; for Wakfu, an estimated 3.5 million players just four years after its official release). From the beginning, Ankama has offered a privileged link to

Dofus players, who can rely on collaborative digital tools to participate in a continuous improvement process for the game. In forums, surveys and contests, players provide frequent feedback, propose improvements, suggest ideas and narrative elements, and vote for new quests or character classes. To do so, they rely on toolkits, which the company uses to keep some control over the creative process and to drive creative efforts toward the contributions it prefers. Dialogue with the community is facilitated by three tools: (1) Ankama's web platform forums, (2) social networks (including Twitter and Facebook) and (3) the AnkaBox (launched in 2011) messaging service that allows players registered in the beta-test to interact with one another, moderators and community managers. Internally, the company tasks 35 full-time community managers with finding ways to meet the expectations of the players.

As the community continued to grow though, the company's internal resources began to be depleted, making it difficult to maintain links with the community (i.e. lack of responsiveness). Initiatives to involve users also became more limited and focused on directional mechanisms (surveys, quizzes, contests) that required less coordination. In response, Ankama chose to outsource some moderating responsibilities to highly experienced players, formalized for certain processes, such as (1) explaining the rules within the games, (2) integrating new players and (3) sanctioning inappropriate behaviour. That is, the outsourcing is partial, but by giving some players more responsibility, it reduces its internal management burdens. These players, by virtue of their status and experience, generally are accepted as moderators by other players. Some previous players with strong gaming experience even were recruited to join Ankama's community manager team. This distributed moderation helps balance out the workload, at least for a time, and enables the company to focus more on developing other communities (e.g. abroad).

For Wakfu, Ankama has sought an innovative management approach. After the four years of beta-testing (2008–2012), it suggested new game operations with regard to how players link up with the environment. The gameplay is the same as available in Dofus (battle mode, quests, guilds, dungeons, character classes), but the management of the environment, natural resources, politics and urban spaces is left to the players. The community thus takes an organic role within the playable environment, as the producer of the game's own dynamics. Players can become citizens and rebuild part of the world destroyed by a previous disaster. They can manage economic activities as craftspeople or traders (innkeeper, blacksmith, farmer) and take on political responsibilities (become mayor of a city, governor of a nation). These actions influence the entire ecosystem. In particular, if they choose a political career, players gain additional rights and duties that they must enact to continue to be supported by other citizens; if they violate those rules, the politicians may suffer sanctions (outlaw status). Other professional careers require players to exploit natural resources with specific life cycles; natural fauna and flora might disappear if players exploit them too much, neglect the soil or consume too many resources. By innovating the conceptualization of the game itself, Ankama thus gives the Wakfu community an empowering opportunity to take responsibility for the game's ecosystem (in its visible and playable manifestation, at least), which creates an appealing entry point for engaged players. The company also adopts a pedagogical approach to educating players, helping them learn to respect the ecosystem with tutorials. In turn, players become more aware of their active, productive role in the game, and Ankama can avoid some of the management demands associated with dynamic game environments.

2. *Key mechanisms of the autonomy–control trade-off*

By empowering players, Ankama orchestrates value logics within its community. Even if the Dofus and Wakfu communities differ, their similar mechanisms led Ankama to adopt parallel orchestration and control choices. For both communities, a primary challenge is entering into dialogue with players (on forums and social networks), to identify and then meet their expectations. The more the community develops, the more expectations the firm faces, and the greater the need for responsiveness. The traditional community management approach, focused on dialogue with users and limits to their derivatives, relies on moderation efforts led by both internal and external community managers, as represented by experienced players. Another important community management issue involves the amount of freedom to give players to improve the artistic universe and products. To orchestrate this core activity, the company uses three mechanisms.

1. Directional (formal, prescribed) and limited in time, such as surveys, quizzes and contests, for which the rules for ideation are defined in advance. This direct management generally implies a specific need of the company, so it solicits players to offer specific contributions in

exchange for rewards, such as in-game bonuses (e.g. potions, weapons). After collecting contributions from many players, the company checks for internal consistency and then selects the ideas to integrate.
2. Indirect, related to ideas submitted organically by players. Ankama's culture of openness allows players to propose improvements and solutions. In both the Dofus and Wakfu game forums (or on social networks), players are free to create posts that launch ideas. If those ideas evoke positive responses in the community, Ankama will study them closely. Community management is critical for assessing players' expectations in this context, particularly in terms of collective acceptance of ideas.
3. Formal and emerging, within the Wakfu game. The professions (fighter, artisan, merchant, citizen, mayor, governor) and associated tools (farm management, trade management, collection and exploitation of natural resources, political career, election system) available to players enable them to manage the game's natural, political and economic ecosystems. Through these devices (technologically locked by tools and socially locked by roles), Ankama guides the creativity and productive dynamics of the game to enrich the ecosystem and avoid deviant behaviours. The goal is for players to regulate their behaviours, both individually and collectively. Players may gain political power, but if they become tyrants, they run the risk of being dismissed or suffer challenges to their authority. Farmers or craftspeople who are too greedy with regard to natural resources (so their behaviours affect other farms) are moderated within the game by other players. For example, the discussion forum features calls for mobilization to restore and rebalance part of the ecosystem in certain areas of the game that have been over-exploited or destroyed. This "regulated autonomy" of the community creates a truly self-organized social, economic and ecological dynamic. Ankama derives two main advantages from these mechanisms: (1) considerable savings of internal resources that otherwise would need to be dedicated to controlling the deviant behaviour of some players and (2) expanded narrative development, because the company is no longer the only one to enrich game dynamics.

These three orchestration mechanisms imply a virtuous logic, insofar as the players are not simply customers but also resources leveraged to develop the company's products. Formal mechanisms technically organize dialogue with the community; social mechanisms encourage moderation (partially outsourced to experienced players) and provoke appropriate contributions from players. More informal levers, supported by the culture of openness and the logic of community functioning, instead involve the delegation of roles and responsibilities, such that players voluntarily engage in more virtuous behaviours that prompt positive dynamics, directed toward the company or game ecosystem. This orchestration model has gradually educated and empowered the Ankama community in its autonomy and behavioural self-management; it is at odds with traditional models used by other massively multiplayer games, with their primary levers of engagement (e.g. fighting, winning, achieving objectives). In the Ankama model, especially in Wakfu, the levers of engagement deal with dynamic ecosystem management at social, political, economic and ecological levels. It is a much more complex, sustainable control system that locks in the community, both technologically and socially. It also constitutes a sort of donation to the community, because players (old and new) gain the opportunity to rebuild and manage the artistic universe created by the company.

IV. Lessons from a Case Comparison

The LEGO Group and Ankama cases demonstrate how companies in creative industries can orchestrate growing user communities to support their innovation efforts. In particular, both examples highlight that a company must continually invest in orchestrating its community to support its evolution and growth. By doing so, it can address the two intertwined tensions: growth vs. appropriation and autonomy vs. control. Different formal and informal orchestration mechanisms emerge, with ex ante or ex post temporal applications, as well as the uses of both push and pull approaches to innovation. A push approach implies that the companies actively engage users in creative behaviours, using formal mechanisms that define the scope of creativity, in terms of its object and temporality. The objective is to stimulate the community to meet a specific innovation need (e.g. idea generation, problem solving). The company uses formal control mechanisms (rewards, evaluation) to ensure users' long-term commitment and to rationalize the results of the creative activity, with a view to ownership. A pull approach instead grants users some form of freedom and creative autonomy, if they already wish to introduce new ideas and address existing problems. In these conditions, companies actively communicate their offer of creative autonomy through informal mechanisms (dialogue, community management, discussion spaces, toolkits) to disseminate the innovation culture more widely. Because they combine both push and pull approaches, the LEGO Group and Ankama effectively orchestrate value creation with their communities, in directive (push) and emerging (pull) ways, and ensure their access to useful, feasible results, while also managing the tension between autonomy versus control.

Furthermore, in both cases, community growth demands new orchestration mechanisms, such as involving users in the management and moderation of the community. With greater community growth, the appropriation of users' contributions may become more difficult, because coordination costs increase, and proximity to users decreases. Thus, new ex post control mechanisms can restore coordination and an alignment with the firm's preferred innovation trajectory. In the LEGO Group and Ankama cases, we find partial outsourcing to expert users as moderators, to reduce coordination costs. They also emphasize user responsibility, which helps reduce mission drift and the costs of moderating efforts.

1. *Ex ante formal control mechanisms*

Ex ante formal control mechanisms are implemented first, because companies proactively seek to collaborate with the community to innovate. Their objective is to manage value creation (creative activity for innovation) with a tool-based, streamlined, pedagogical process that guides users toward value-creating behaviours and creativity that will support the desired goal. The company must think about how to formalize and control this activity before initiating the collaboration with the community. In particular, it should define three parameters: (1) the innovation need, to communicate the objective and purpose to the community; (2) the tool to be used (competition, game, challenges, problem solving, survey), which establishes the technological and pedagogical means to promote creativity according to the desired purpose and (3) the temporality, so that the creative action refers to a specific period and can quickly involve users. From this mechanized perspective, toolkits have an important role; they provide a technological framework for users to innovate, regardless of the community growth rate. Then as the community grows, the number of users who can create value increases. However, within a streamlined creative process, toolkits also slow down or limit users' creativity. In the LEGO

Group case, for example, the firm first provided a specific toolkit to frame the ideas suggested by users, before allowing various toolkits to grant more flexibility to the growing community and avoid constraining users' creativity.

2. *Ex post formal control mechanisms*

Formal ex post mechanisms control the results of users' creative behaviours to facilitate their appropriation. The objective is to control value creation with internal (managed by the company) or external (managed by the community) evaluation processes that ensure consistency with the business strategy, innovation needs and feasibility. Formalized evaluations depend on the tool used to stimulate creativity. For example, in an ideation competition, formal selection rules and criteria can encourage users to create and propose elements that match the company's innovation needs. This formalization helps ensure the results are meaningful, in that the users can understand (1) how their ideas or creations will be evaluated by the community or company and (2) which characteristics their ideas must possess to be eligible and useful. Recognition and remuneration mechanisms can maximize this alignment; the LEGO Group's recognition efforts align with its formal guidelines, rules and expectations. In its formalization process, it also has shifted incentives for successful contributors, such that the main reward highlighted on the platform is no longer a 1% royalty but instead enhanced status, "glory and recognition (be a star and sign autographs for your fans at the official LEGO Ideas signing event)."

3. *Ex post informal control mechanisms*

Informal mechanisms are crucial for encouraging persistent innovation with a community, reflecting their more social approach, rather than a formal, technical, focus on framing, organizing, stimulating and rationalizing creative processes. In particular, they can help manage the community's growth through outsourcing. As any community grows, its coordination costs explode, and companies must find ways to absorb them. The LEGO Group and Ankama cases feature two mechanisms: (1) moderation by expert users, who can take charge of welcoming newcomers, explaining the community's operating rules, or sanctioning classic abuses, and (2) education about respectful social behaviour, dissemination of good conduct rules and reminding users of the rules so that they can work together. By extending community governance in this way, the firms build trust and also remind users that they are responsible for the common social system. These moderators thus need to be expert users, because social recognition of expert status, by both company and the community, grants these users legitimacy in their educational and

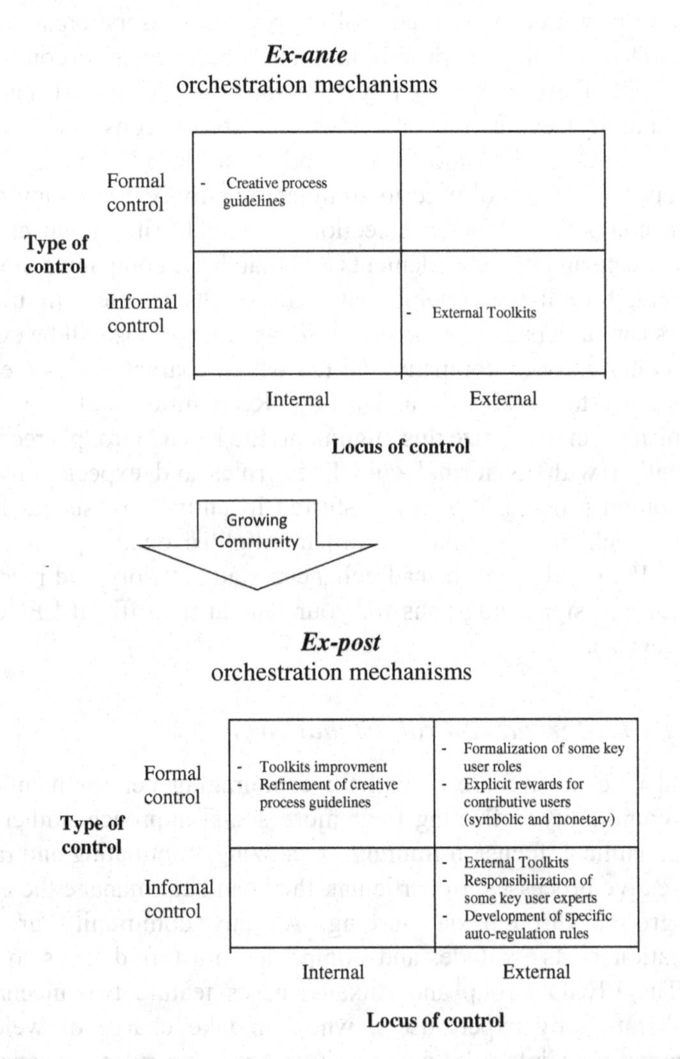

Figure 2: *Ex ante* and *ex post* user community orchestration mechanisms.

moderating activities. Therefore, they can help manage tensions between community growth and control, by involving community members.

However, such partial outsourcing might not be enough to manage the tension between user autonomy and appropriations of their contributions. As noted, growing communities require greater control but still must retain enough autonomy to foster creation and innovation. Companies therefore need another informal mechanism, in the form of voluntary roles and responsibilities to structure the community. Users take on specific functions, with a defined scope of action and specified degree of autonomy. The roles are structuring; they make users accountable to the company and support better organization of collaborations and transactions among members. The complementary roles actually structure the community and make users aware of their complementarity. Such efforts enable autonomy management, because autonomy depends on the roles that users themselves choose. For example, in Wakfu, a role choice implies a realistic business function that determines the user's behaviour, freedom of action and autonomy, along with the game's ecosystem. Players thus empower themselves, in terms of behaviour and creativity. As producers of part of the game dynamics, they are responsible for its evolution. Their creations are directly and easily appropriated by the company, because the chosen roles guide their creative activities and automatically make them eligible. With this principle, the company empowers the community, reduces its control costs and increases the level of ownership.

Ultimately, informal mechanisms affect the social structure of the community, through the recognition of a hierarchy (acceptance of moderating role by expert users) and the empowerment of community members (development of respectful behaviour, self-regulation, definition of role and responsibility).

V. Conclusion and Implications

Previous, scarce literature on community orchestration implicitly suggests that control of external communities requires formal mechanisms, like contracts. But as these two successful cases show, managing external communities is more tricky than that, so firms need to attend to the social welfare and interests of the community. The type, time, space and level of control are meaningful determinants of community dynamics and increase the level of ownership (Figure 2). Complex orchestration mechanisms can help the firm deal with the evolution of the community; this evolutionary perspective on community suggests, contrary to some recent assertions (Parker and Van Alstine, 2018), that control mechanisms are not stable over time. Firm orchestration efforts thus should pursue ad hoc, external control mechanisms compatible with the community's evolution, rather than establish a seemingly permanent, optimal control scheme. To keep pace with community growth and cope with users' variety, Ankama and the LEGO Group adjust and expand the control space, without hampering — or more accurately, by empowering — autonomy.

Therefore, to respond to fast growing communities, firm-hosted user communities in creative industries should use carefully designed, specific orchestration mechanisms. As the community grows, they should complement formal ex ante control mechanisms with formal and informal, external control mechanisms. This morphology of control, from internal firm rules to more pervasive external regulation mechanisms, can respond to increases in both the number and types of contributions, without disrupting a strategic orientation toward innovation or encouraging deviant behaviours. A key orchestration mechanism for a growing community thus is being flexible, along with enriching social coordination and support between the firm and the community.

Our findings still should be interpreted with caution though. Ankama and the LEGO Group are atypical firms, in that they have accumulated extensive experience with external communities, so they already have achieved substantial maturity in their orchestration efforts. This caution further implies that informal, external behaviour control mechanisms may require multiple trial-and-error cycles. We in turn deduce that these processes are idiosyncratic in nature and would be difficult to replicate in other contexts, which naturally feature different competences. Therefore, this study provides the insight that orchestrating the growth of external communities is conditioned by the coevolution of internal and external behaviours, which cannot be predetermined but can be effectively addressed with the dynamic capabilities of the firm.

References

Boudreau, K. (2010). Open platform strategies and innovation: Granting access vs. devolving control. *Management Science*, *56*(10), 1849–1872.

Burger-Helmchen, T. and Cohendet, P. (2011). User communities and social software in the video game industry. *Long Range Planning*, *44*(5–6), 317–343.

Dahlander, L. and Magnusson, M. (2008). How do firms make use of open source communities? *Long Range Planning*, *41*(6), 629–649.

Dhanaraj, C. Parkhe (2006). A: Orchestrating innovation networks. *Academy of Management Review*, *31*(3), 659–669.

Jeppesen, L. B. and Frederiksen, L. (2006). Why do users contribute to firm-hosted user communities? The case of computer-controlled music instruments. *Organization Science*, *17*(1), 45–63.

Katz, M. L. and Shapiro, C. (1985). Network externalities, competition, and compatibility. *American Economic Review*, *75*(3), 424–440.

Laursen, K. and Salter, A. (2005). Open for innovation: The role of openness in explaining innovation performance among UK manufacturing firms. *Strategic Management Journal*, *27*(2), 131–150.

Lave, J. and Wenger, E. (1991). *Situated learning: Legitimate peripheral participation*. Cambridge University Press.

Parker, G. and Van Alstyne, M. (2018). Innovation, Openness, and Platform Control. *Management Science*, *64*(7), 3015–3032.

Piezunka, H. and Dahlander, L. (2015). Distant search, narrow attention: How crowding alters organizations' filtering of suggestions in crowdsourcing. *Academy of Management Journal*, *58*(3), 856–880.

Piller, F., Ihl, C., and Vossen, A. (2010). Customer co-creation: Open innovation with customers. *Wittke, V./Hanekop, H*, 31–63.

Sarazin, B., Cohendet, P., and Simon, L. (2017). Les communautés d'innovation. *Editions EMS*.

Shankar, V., and Bayus, B. L. (2003). Network effects and competition: An empirical analysis of the home video game industry. *Strategic Management Journal*, *24*(4), 375–384.

Von Hippel, E. (1986). Lead users: A source of novel product concepts. *Management Science*, *32*(7), 791–805.

von Hippel, E. and Katz, R. (2002). Shifting innovation to users via toolkits. *Management Science*, *48*(7), 821–833.

West, J. (2003). How open is open enough? Melding proprietary and open source platform strategies. *Research Policy*, *32*(7), 1259–1285.

© 2021 World Scientific Publishing Company
https://doi.org/10.1142/9789811234286_0013

Chapter 13

Crisis Communities: New Forms of Action During the COVID-19 Health Crisis

Zoé Masson and Guy Parmentier

Crisis is an exceptional situation that creates a break in continuity with usual activities (Cros *et al.*, 2019) in which the social system cannot solve with the usual solutions the problems that are necessary for its sustainability (Habermas, 1975). It challenges people's view of reality (Weick, 1988), creates uncertainty about the future and is thus a major stressor. The COVID-19 crisis has a characteristic of its own, the rupture was both spatial and temporal. Overnight, millions of people found themselves confined to restricted spaces. For many of them, the time available for personal activities increased dramatically, creating a paradoxical situation of spatial compression and temporal expansion. During this crisis, Internet consumption has increased sharply (Bourdeau-Lepage, 2020), to communicate by videoconference, to search for information or to connect to groups of friends on social networks. *Being alone together,* during confinement, individuals mobilized online communities to support and entertain themselves, and to contain a virus on a global scale. Crises push individuals to improvize to find solutions by developing multiple interactions (Adrot and Garreau, 2010). In this situation, individuals mobilize their creativity (Drazin *et al.*, 1999) and virtual communities are a favourable place for sharing the creativity of Internet users (Parmentier, 2015).

Sociologist Ferdinand Tönnies originally defined communities as *"collectives based on geographical and emotional proximity, and involving direct, concrete, authentic interactions between its members"* (Proulx and Latzko-Toth, 2000, p. 101). With the development of the Internet, these communities also exist online by developing a new form of socialization conducive to knowledge sharing and creativity (Dahlander *et al.*, 2008; Parmentier, 2015). Thus, in response to the problems caused by confinement, new communities have emerged in many fields such as education, sports, politics, local life, research, engineering, etc. These communities seem to have their own characteristics that distinguish them from the traditional categories of virtual communities identified in the literature. Understanding these communities could allow us to draw lessons for dealing with possible future health crises. Thus, from the testimony of 7 community creators and managers (see Appendix 1), we deduce the characteristics of these communities and reveal elements of community management that could be applied in a crisis context and in a context of normality.

I. A Societal Motivation

The creation of online communities is motivated by an interest in exchanging among interlocutors around a common subject, often centred around specific issues. Members are involved, seeking personal benefit, and the more they feel the contributions are of high quality, the more they feel committed and contribute in turn (Wiertz and de Ruyter, 2007). In this health emergency, community building is spontaneous: initiatives seek to respond to an immediate, a priori ephemeral need. Projects are often impulsive in order to meet new needs: the need to find health solutions, to help one's relatives or to prepare to face loneliness in confinement whose end point is unknown. This is how Mehdi recounts the birth of the French community initiative *Les 10 minutes du peuple*: a Facebook group encouraging all its members to meet every night from the confinement at 7:30 pm to dance together, in front of their camera. The idea is born when he meets his roommate *He comes to talk to me and says: what do you think, we'll make all of Paris dance at the same time.* A short while later, in Niort, Matthieu was thinking over his coffee about a way to help the people in his town for whom confinement would be complicated. That same evening, he had his project *Sans sortir* ready to go. As for the

Hacklacrise project initiated by Sébastien, it was born from an impulse on the social network Linkedin, when he shared a video encouraging to organize a hackathon to help people in difficulty *I had written three–four keywords, but otherwise it was really crying from the heart.*

When impetus is the driving force behind such initiatives, their objectives are rarely explicit. The sudden birth of these virtual gatherings is driven by a trend, without really knowing in which direction to move in order to make the initiative effective: *Our 10 Minutes du peuple community does not have a well-targeted objective* (Mehdi). The reason of being of these communities seems to take shape over the course of their existence, initially responding to very broad values, which can encompass a variety of contents: *The vocation of the lab is to have a positive impact on society* (Myriam). Humour thus takes pride of place in the anxiety-provoking climate imposed by the health situation, and crisis communities also become places of relaxation for everyone: *What was really pleasing, even for me, was that we were really stressed during this period, it was a time for decompression on Saturdays during the challenge* (Bertrand).

The motivation for individuals to get involved in these communities suddenly becomes more societal. The crisis context has transformed online communities, which are now based on values such as mutual aid, solidarity and entertainment: *I make it a point of honour that there really is this side of solidarity exchange* (Matthieu). This transformation is consistent with the fact that individuals care more about their relatives in this period of health crisis (Bourdeau-Lepage, 2020). Moreover, these communities are also a way for Internet users to support each other morally in this period of crisis. Social support is a significant help in combating the negative effects of confinement, and staying connected to others, if not physically, atleast virtually, is an undeniable support (Saeri *et al.*, 2020). A 2020 study on engagement in online communities during the lockdown in England and Ireland shows that 70% of individuals engage to offer support to other users. They also find that doing so makes them feel calmer afterwards and less anxious (Elphick *et al.*, 2020).

II. A Large Opening

The profiles of Internet users invested in an online community often share common points: an attraction for a discipline, an issue or values. Virtualization allows these individuals to share their knowledge and ideas despite geographical constraints, without knowing each other "physically", and collective emotions are born between individuals (Salmela, 2014) and the desire for a physical encounter sometimes ends up intervening in places of virtual exchanges (Parmentier et Gandia, 2013). Collective emotion based on strong bonds and rituals sometimes makes access to the community difficult for neophytes who can easily feel rejected (Proulx, 2006). In the context of the health crisis linked to COVID-19, the motivations that led to the creation of new communities are more related to the common cause than to specific problems. Access to them is facilitated and the constraint of legitimacy to commit to them is erased by the desire to rally around values of mutual aid and support.

The moderators of the crisis communities are the first to point out that these virtual spaces are open to everyone: *We didn't think there would be so many people. Today there's everyone on the 10 minutes, there are people who are fifty, there are people who are seventy, there are people who are students, there are dancers, artists, musicians* (Mehdi).

Moreover, this global crisis seems to increase tenfold the desire for mutual aid, leading communities to communicate on unifying values. Communities can be places of sharing and mutual aid or effective means of action to combat a health crisis of this magnitude. In both situations, the spirit is one of collaboration: *The idea is to say to each other "let's try to learn from this crisis, to find new uses, to be more open than before and to federate people"* (Myriam).

Internet users on these crisis communities also take advantage of the strong virtualization effect caused by confinement to make encounters that might not otherwise have taken place. This characteristic already identified within traditional virtual communities (Dahlander *et al.*, 2008) also applies in the case of communities born in times of crises. The emergence of a geographical *melting pot* is confirmed by moderators. At the national level, the French seem to merge between city dwellers and rural dwellers: *They are tech people, but in an extremely broad sense, it goes from the Parisian marketer to the guy who was like fifty, bearded, who lives in the Morvan and who is the kind of guy who learned tech on the job* (Sébastien). Internationally, communities are growing by bringing together Internet users from all continents: *We've got listeners from all over Latin America, North America, a few countries in Africa and Australia* (Mehdi), *Representatives from over 60 countries have registered, from every continent!* (Bertrand).

This openness opens up new ways and opportunities to work together. This mix of individuals with different profiles has, for example, enabled Sébastien to run a unique *hackathon.* Accustomed to organizing these events during physical meetings, digitalization has allowed him to bring together participants who were usually interested, but had difficulty travelling: *There you have teams from all over France working together, it's very interesting* (Sébastien).

Another form of openness manifests itself in the sharing of the platform that hosts the community. The inventor of such a virtual platform makes a significant effort to implement the community, involving financial and time costs. The support is often unique, and if it were to be shared, would certainly imply a financial arrangement. In the case of communities born in this period of crisis, creators were quite open to the diffusion of their platform, without any compensation, except to offer a wider range of people the possibility of using it. Matthieu was already thinking about this aspect when he first sketched out his concept: *I thought of it so that it could be appropriated, so that if the initiative resonates with certain communities and others, I could share it and it could grow and be appropriated by everyone* (Matthieu).

III. An Acceleration of Community Building

Confinement was set up in an accelerated manner and within a few days the French were not allowed to leave their homes. This phenomenon of acceleration was transferred to the web where, on the one hand, the moderators already had the technological tools necessary to create new virtual spaces and, on the other hand, Internet users responded overwhelmingly in record time. Technologically, the rapid emergence of crisis communities can be explained in part by the use of pre-existing tools. They allowed low development costs and, above all, rapid start-up in this emergency period. Social networks, for example, have enabled many Internet users to bring together hundreds of people in a wide variety of groups. Among the seven interlocutors contacted for this chapter, two chose to use Facebook (Bertrand with the *Confined Sport Challenge* and Mehdi with *Les 10 minutes du peuple)*. As for Sébastien, he chose to use the free collaborative communication platform *Slack.* He was able to bring together nearly 1,000 people in less than a week by communicating exclusively on the Linkedin social network. Others relied on pre-existing content to which they added a "COVID-19" space. This enabled them to benefit from an operational site and a pre-existing community offering numerous tools and means of exchange. This is the case of the *3D experience Lab* community which has existed for five years and which created an *Open Covid* space at the beginning of March, when the confinement started: *The idea of the creation of the Open Covid 19 community which was created in 2h, well in 1h, well in 3 clicks* (Myriam).

Bathed in an emergency situation, Internet users rushed into a kind of "rush to virtual communities". This massive, rapid and unexpected craze surprised the moderators: *In almost 24 or 48 hours, there were 500,000 subscribers, we didn't think it was going to happen* (Mehdi). Mathias, from the video game studio *Celsius Online* confirms this: *It was a wave, we increased the number of people connected by 700%* (Mathias). It's the same testimony when Sébastien talks about the launch of his project: *I knew I was going to talk to people, then I didn't expect so many of us* (Sébastien). Media communication has also played a catalytic role in the spread of these communities, often reported by the French press and media seeking to highlight the opportunities for action and entertainment available to individuals in times of confinement.

Every stage in the life of these communities seems to be accelerating. While the launch and affluence phases are getting up to speed more quickly, so too is the production of ideas and the launch of concrete projects. The speed of action and results is an obvious observation: *We went superfast, it's never been seen before (Laurent)*. In the *Open Covid* community, there were three times more ideas in 2 months than in the other public communities of *3D Experience Lab* in five years, and out of 138 ideas, 16 were already in the final phase of realization.

IV. A Place of Creativity

Not all virtual communities are creative from the outset. Crises have an effect on creativity. For example, in large technological development projects, the creativity of employees has a major role to play in solving major problems that arise during periods of crises (Drazin *et al.*, 1999). In addition, improvization is a response to crises in organizations (Barrett, 1998), which in the digital world can be akin to creative DIY (Rüling and Duymedjian, 2014). We note that certain virtual communities that appeared during the crisis are from the outset focused on creativity, such as *Hacklacrise, Open Covid* or *Sans sortir* in which the creator has organized creative challenges: *The idea is to encourage creation, to encourage initiative* (Matthieu).

In other communities that did not organize contests or a section dedicated to creation, creativity has nevertheless developed strongly. The strong motivation of the participants, the numerous contributions and the openness of these communities probably fostered collective creativity in the community.

As we have seen earlier, the sense of urgency related to the crisis has created a strong motivation to participate in these communities. Motivation is one of the basic elements of individual creativity (Amabile, 1997) and of creativity in user and brand communities (Parmentier, 2015). This sense of urgency pushes individuals to engage strongly in creative activities: *Since there was a sense of urgency and it was a matter of life and death, it probably also mobilizes minds in a more constructive way... There was also this need, of course, to find solutions very quickly* (Myriam). In an online community, there is a strong correlation between its demographic growth and that of its content (Roth *et al.*, 2008). During this crisis, in communities that experienced strong growth, content also experienced strong growth, as the manager of the online *Renaissance Kingdom* community notes: *the increase in the number of posts was greater than the increase in the number of players* (Mathias). With more user productions, statistically, it is possible to envisage an increase in creative ideas within these crisis communities, as the innovation manager for the *Open Covid* community notes: *A huge number of ideas at DS: [he talks about the ideas submitted] there have been a total of about 1000 over 5 years, so you see almost 150 over two months, which is huge compared to what we were used to seeing* (Myriam). Even in communities that are not dedicated to innovation or creativity, we see the emergence of unanticipated creativity: *We thought we were sharing a handful of activities and very quickly it multiplied incredibly. Our members through their own challenges brought a lot of creativity to everyone* (Bertrand). The diversity of participants is a factor of creativity in online communities (De Toni *et al.*, 2012). Thus, the opening of the crisis communities to multiple profiles of participants seems to be one of the factors that fostered their creativity: *Innovation came with the meeting of people who were from different worlds: from the medical world with makers, with engineers, with companies. That was the key to success* (Myriam).

In the end, this creativity did not remain without tangible results. The Confided Sport Challenge community produced a digital book of testimony at the end of the confinement: Ultra confined, and the Open Covid and Hacklacrise communities have allowed the emergence of operational projects: *There are 2–3 projects that have really worked well* (Myriam), *The most successful is the home class and SOS equipment* (Sébastien), *it also allowed us to identify projects that are now in operation* (Myriam).

V. A Lasting Phenomenon?

After two isolated months, individuals emancipate themselves from their virtual universe to return little by little to the "real" world. The fight against the virus is running out of steam and the dynamism of crisis communities seems to be dying out at the same rate. But it is complicated to extricate oneself so quickly from a community that has been carried through this period of crisis, with real consequences for individuals.

For some people and organizations, the craze for these online platforms has had a stimulating effect, awakening projects and desires that had sometimes been imagined for a long time. This is the case at the *Celsius Online game* studio *for us it precipitated the decision to create a mobile game* (Mathias) where a mobile application has been in development for two months. For others, it opens up new ways of working. Within the *3D experience Lab* is developing the *Fast Track*, a new tool to speed up the selection process of a promising startup, usually too long to make it efficient in times of emergency.

The end of this period of confinement and the gradual resolution of the health crisis sounded the death knell for the communities created for the occasion: *Now that's it, there's no need for us anymore, the projects no longer really need to exist, they were projects that really responded to the crisis* (Sébastien). However, the moderators and users were aware that these projects were responding to an ephemeral need: *It's not embarrassing that the group is slowly dying, is a collective meant to be sustainable? I don't think so, otherwise you direct your actions to try to maintain it* (Laurent). However, letting the communities built during this period run out of steam is complicated for some individuals who find it hard to stop: *I have a twinge in my heart, I don't really know what to do* (Bertrand). New solutions are then considered, and some have given in to the infatuation created by their community to think about a coherent and adapted transformation into a group that would make sense to last over time: *In the future, maybe we will do an event once a month or every two months to bring people together* (Mehdi). A strong will to meet users was also born in many of these online groups: *When things calm down, we'll have a beer together, and then we'll meet each other because they're people, I've never seen them, I can't wait to see them in real life* (Sébastien).

VI. Discussion and Conclusion

By studying new communities that emerged during the crisis, we have drawn some of the characteristics of these crisis communities based on improvization, societal motivation, openness to the greatest number and diversity of participants, a rapid rate of growth, great creativity in productions, and a temptation to continue the adventure beyond the crisis. The spatiotemporal upheaval induced by this crisis, a double movement of spatial compression and temporal expansion, pushed the confined people to turn to online tools to continue working, socializing and occupying their free time. During this crisis, the virtualization of human activity has shown a double face of opening up possibilities and reducing the feasible, echoing two initial conceptions of the virtual identified by Proulx (2000) in the social field: the resolution of a world stricken with imperfection or the false approximation of a reality too complex to be simulated. Crisis communities are more in line with the third conception of Proulx (2000), which considers the virtual as a hybridization of the real and the virtual. In a crisis situation, as Levy (1998) indicates in his essay on *qu'est-ce que le virtuel (what is virtual)*, the virtual reopens the possibilities, blurs the established distinctions and increases the degrees of freedom. The "virtual" nature of the community is thus an interesting tool to fight against the crisis by bringing new ideas and new forms of action. The context of a health crisis is thus a proven catalyst, but it also highlights certain elements that could weigh in the management of virtual communities during a crisis or in a context of normality. The study of the motivations of moderators and users shows that displaying a cause and societal values attracts a more massive panel of users. This echoes the fact that an openness to all profiles seems to avoid the brake of illegitimacy to integrate too specific a community. The temporality in these online communities gives a fundamental rhythm to their activity. Posting *deadlines* and projects in a short timeframe seems to stimulate participants and encourages them to visit the community on a more regular basis. The creative activity flows from the previous observations through a larger number of contributions. Also, adding a space for challenges and contests seems to contribute to the creative excitement of individuals. Finally, these crisis communities highlight the major role that

Appendix 1: Summary of empirical material.

Code Name	Function	Community	Theme
Myriam	Partnership Manager	3D experience Lab	Industry
Laurent	Participant	Profs-Chercheurs	Education
Bertrand	*Community Manager*	Confined Sport Challenge	Sport
Sébastien	*Community Manager*	Hachlacrise	Health
Mathias	Production Manager	Loup Garou en ligne	Video game
Mehdi	*Community Manager*	Les 10 minutes du peuple	Entertainment
Matthieu	*Community Manager*	Sans sortir	Local initiatives

virtual communities can play in maintaining social ties and in the collective search for a collective solution to a major problem.

References

Adrot, A. and Garreau, L. (2010). Interagir pour improviser en situation de crise. *Revue Francaise De Gestion, 203*(4), 119–131.

Amabile, T. M. (1997). Motivating creativity in organizations: On doing what you love and loving what you do. *California Management Review, 40*(1), 39.

Barrett, F. J. (1998). Creativity and improvisation in Jazz and organizations: Implications for organizational learning. *Organization Science, 9*(5), 605–622.

Bourdeau-Lepage, L. (2020). Le confinement et ses effets sur le quotidien, Premiers résultats bruts des 2e & 3e semaines de confinement en France. Université Jean Moulin Lyon 3.

Cros, S., Lombardot, E., and Real, B. (2019). Manager sous stress aigu en situation de crise. *Revue Francaise De Gestion, 282*(5), 37–56.

Dahlander, L., Frederiksen, L., and Rullani, F. (2008). Online communities and open innovation: Governance and symbolic value creation. *Industry & Innovation, 15*(2), 115.

De Toni, A. F., Biotto, G., and Battistella, C. (2012). Organizational design drivers to enable emergent creativity in web-based communities. *Learning Organization, 19*(4), 337–351.

Drazin, R., Glynn, M. A., and Kazanjian, R. K. (1999). Multilevel theorizing about creativity in organizations: A sensemaking perspective. *Academy of Management Review, 24*(2), 286–307.

Elphick, C. E., Stuart, A., Philpot, R., Walkington, Z., Frumkin, L. A., Zhang, M., Levine, M., Price, B. A., Pike, G., Nuseibeh, B., and Bandara, A. K. (2020). Altruism and anxiety: Engagement with online community support initiatives (OCSIs) during Covid-19 lockdown in the UK and Ireland. *Computer and Society*.

Habermas, J. (1975). "Legitimation crisis," in Beacon Press, Boston.

Levy, P. (1998). *Qu'est ce que le virtuel?* La Découverte, Paris.

Parmentier, G. and Gandia, R. (2013). Managing sustainable innovation with a user community toolkit: The case of the video game Trackmania. *Creativity and Innovation Management*, *22*(2), 195–208.

Parmentier, G. (2015). How to innovate with a brand community. *Journal of Engineering & Technology Management*, *37*, 78–89.

Proulx, S. (2006). "Communautés virtuelles: ce qui fait lien," *Presses de l'Université Laval,* pp. 13–26.

Proulx, S. and Latzko-Toth, G. (2000). La virtualité comme catégorie pour penser le social: L'usage de la notion de communauté virtuelle. *Sociologie et société*, *32*(2), 99–122.

Roth, C., Taraborelli, D., and Gilbert, N. (2008). Démographie des communautés en ligne. *Reseaux*, *152*(6), 205–240.

Rüling, C.-C. and Duymedjian, R. (2014). Digital bricolage: Resources and coordination in the production of digital visual effects. *Technological Forecasting and Social Change*, *83*(0), 98–110.

Saeri, A., Greenaway, K. H., and Cruwys, T. (2020). Why are we calling it "social distancing"? Right now, we need social connections more than ever. *The Conversation*.

Salmela, M. (2014). The Functions of Collective Emotions in Social Groups, dans Konzelmann Ziv A., Schmid H.B. (dirs.), *Institutions, Emotions, and Group Agents*, Springer Netherlands, Dordrecht, 159–176.

Weick, K. E. (1988). Enacted Sensemaking in Crisis Situations. *Journal of Management Studies*, *25*(4), 305–317.

Wiertz, C., and Ruyter, K. de (2007). Beyond the Call of Duty: Why Customers Contribute to Firm-hosted Commercial Online Communities. *Organization Studies (01708406)*, *28*(3), 347–376.

© 2021 World Scientific Publishing Company
https://doi.org/10.1142/9789811234286_0014

Chapter 14

Communities of Innovation: From Co-Creation to Resilience

Madanmohan Rao

This chapter reviews some of the literature on knowledge management, innovation, entrepreneurship and resilience, and shows how communities of practitioners can be key players in this regard. Based on insights from 25 organizations, it highlights emerging trends in the field of innovation communities, such as the growth of cross-sectoral and cross-organizational communities. During the pandemic crisis and in the post-COVID era, resilient communities will play a key role in sustaining and scaling the next waves of innovation.

Based on insights from Communities of Practice (CoPs) for Knowledge Management (KM) and innovation in 25 organizations, this chapter identifies three trends in the field. Innovation is becoming increasingly cross-disciplinary, and calls for CoPs that cut across departments and disciplines. The growing pace of innovation across industries calls for CoPs that are inter-organizational, and bring in valuable outside perspectives. And finally, the coronavirus crisis is calling for CoPs that strengthen resilience of operations and innovations.

Two other drivers of innovation are noticeable from these case studies. One is the scale and speed of digital transformation, which is being hastened even more due to the COVID-19 pandemic and its requirements for work from home (WFH), thus driving more CoPs to operate in virtual mode. The second driver is the accelerator model of innovation, where companies are increasing the scope and quality of innovations by partnering with startups in a structured manner to tackle emerging opportunities and challenges. This calls for communities that can absorb diverse cultures, ranging from agile entrepreneurs to established incumbents. With the world's third largest base of startups, India is an attractive location for corporates to harness entrepreneurs for innovation via co-creation, as many of these case studies illustrate.

I. Methodology

In the period 2019–2020, Bangalore-based heads of programs related to innovation and knowledge management in seven organizations were contacted to submit case study profiles exclusively for this chapter. A 20-question survey on innovation Communities of Practice (CoPs) was sent via email to the business heads, divided into three sections: background (e.g. CoP scope, values, membership), activities (e.g. interactions, incentives, outcomes) and trends (e.g. technology, behaviors, COVID-19 impacts).

Responses were collected and curated over two or three rounds of online communication, phone calls and meetings. Edited versions of these case profiles are presented in this chapter. Other articles related to their innovation activities were also published in other media and shared on professional social media platforms like LinkedIn. Additional insights have been added from references to these organizations in the literature review, and conversations with these business heads during industry conferences, seminars and workshops. These seven organizations are: Bosch, Target, Trianz, Brigade Group, NetApp, Unisys and The Indus Entrepreneurs (TiE).

CoP profiles of seven other organizations are drawn from the Global Virtual Knowledge Summit 2020, featuring winners of the Most Innovative Knowledge Enterprise (MIKE) awards. Heads of knowledge management and innovation programs from seven organizations were interviewed during July-August 2020 for more insights into the functioning and impact of CoPs: EY, Tata Chemicals, Petroleum Development Oman, BINUS University (Indonesia), Mobarakeh Steel Company (Iran), Cognizant Technology Solutions and Afcons Infrastructure (all interview links are listed in the reference section).

II. Literature Review

CoPs constitute a growing part of the organizational landscape of 21st century organizations. Effective CoPs can improve productivity, employee engagement, knowledge continuity, innovation, risk management and resilience. They call for careful nourishment and sustenance, with clear commitment to value and sustainability. Over the years, CoPs have come to be known by various catchy names (Rao, 2004) — they include Learning Networks (in HP), Best Practice Teams (Chevron), Family Groups (Xerox), COINS (Ernst & Young's community of interest networks) and Thematic Groups (World Bank). Corporate yellow pages have been known variously as PeopleNet (Texaco) and Connect (BP).

Online CoPs are emerging as a powerful tool for knowledge exchange and retention. APQC classified CoPs into four types: helping (peer-to-peer sharing of insights, e.g. Schlumberger's Eureka, DaimlerChrysler's TechClubs), best-practice sharing (sharing of documented verified user practices, e.g. Schlumberger's InTouch), knowledge-sharing (connecting of members, e.g. CGEY) and innovation (cross-boundary idea generation, e.g. Siemens ShareNET).

Typical roles in organizational networks include central people, peripheral people, boundary spanners and knowledge brokers. "Network structure can facilitate or impede effectiveness of knowledge workers," according to Cross (2004). Social network analysis (SNA) in CoPs has implications for organizational leadership, social ecology, relational development and network planning. Beerli, Falk and Diemers (2003) classified such organizational networks into the following types: experiencing network (e.g. Seven Eleven Japan), materializing network (e.g. Sharp), systematizing network (e.g. Accenture) and learning network (e.g. Buckman Labs).

Businesses today are driven as much by projects and portfolios as by processes, due to increasing 'projectization' in shorter business cycles. "In the projectized organization, strategy has become the delivery of a series of aligned projects," according to Shelley (2017). This calls for effective community sharing of knowledge across projects, which needs continuous upgradation of skills in conversation, listening, questioning, reflection, facilitation and use of digital tools. Organizations need to combine leadership (exploration, choice of an optimal path) and management (prioritization, delivery) in their knowledge journeys.

During times of crisis, it is key for companies to retain critical knowledge when employees are being fired or laid off, or if they leave on their own accord. While CoPs can help retain some of this knowledge, additional mechanisms like exit interviews, critical incident reviews and accelerated apprenticeship are called for (Leonard *et al.*, 2014). Organizational resilience during crises calls for robustness of operations, strategy, culture and leadership (Välikangas, 2010). Companies need to be able to overcome the bias of past success and the fear of failure to move on and bounce back from a crisis.

While the bulk of knowledge management and innovation literature has focused on the business sector, community sharing of knowledge is important in government, academia and civil society as well. For example, Singapore's Information & Knowledge Management Society (iKMS) has organized annual awards for excellence in knowledge management, which were won by 12 government agencies (Tan and Rao, 2013). A focus on leadership, process, technology and community engagement helped deliver benefits in productivity, compliance and growth.

A core part of the innovation capacity of cities and nations comes from creative and entrepreneurial communities. Rosenberg (2002) analyzed how "Silicon Valley clones" or clusters of knowledge industries are emerging and faring in Cambridge, Bangalore, Singapore, Helsinki, Tel Aviv and Hsinchu. Success factors include business webs, local "living laboratories," entrepreneurship support organizations, local academic and research institutes, and commercial partnerships between academia and industry.

Strengthening the innovation quotient of local entrepreneurial ecosystems calls for community approaches to multiplying connections among entrepreneurs and mentors, improving access to entrepreneurial education and creating events and activities that activate all the participants in the startup community (Feld, 2012). In the era of the entrepreneur, communities of startups hold a key to rejuvenating the economies of many cities around the world as they create the basis for the next wave of economic growth.

Community dynamics in creative clusters helps the growth of technology, business and cultural innovators (Gannett, 2018). The number, quality and duration of relationships in such communities lead to varying interactions with mentors, collaborators, muses and promoters. Creative density has knowledge spillovers, via formal and informal as well as planned and serendipitous encounters. Stability and a conservative focus may come from large incumbent organizations, but novelty and disruptive ideas come from the periphery.

In a fast-moving global economy where customers are more digitally connected, informed and empowered than ever before, communities of co-creation with partners and customers may work best for organizational success (Jansen and Pieters, 2018). Continuous contact with customers via community forums helps unearth insights about unmet needs much faster, and paves the way to complete co-creation and even co-ownership.

III. CoPs for KM, Innovation and Resilience: MIKE Award Winner Profiles

Organized by a research group led by Hong Kong Polytechnic University, the Most Innovative Knowledge Enterprise (MIKE) awards build on the earlier Most Admired Knowledge Enterprise (MAKE) award, launched by Rory Chase of Teleos in 1998. The Global MIKE Study Group has developed a framework of eight parameters for winner scores — empowerment of knowledge workers, transformative leadership, user experience, knowledge networks, innovative culture, knowledge-based offerings, knowledge creation processes and creative spaces. Together, these criteria cover human capital, relational capital, innovation capital and process capital. CoPs play an important role in this regard.

Eight countries and regions took part in the 2019 Global MIKE study: China, Hong Kong SAR, India, Indonesia, Iran, Japan, the Middle East and the US. Seven winners of the 2019 awards were featured at the Global Virtual Knowledge Summit 2020, organized by the Confederation of Indian Industry as a virtual event on July 6–8, 2020. The winners were approached for in-depth interviews, with key CoP highlights summarized in Table 1 and described in greater detail in what follows.

1. *Cognizant Technology Solutions*

Digital transformation services firm Cognizant has established CoPs at three broad levels. There are hundreds of internal CoPs initiated by various business units and teams to enable the exchange of knowledge and find resolutions to problems. Some teams leverage such communities to crowdsource ideas for a specific business or customer needs. Partnership

Table 1: CoPs at MIKE award winner organizations.

Organization	CoP highlights	Overall impacts of KM and innovation
Cognizant	Hundreds of CoPs. Three types of CoPs: internal, partner (forums, hackathons), customer (share success stories).	Individuals (continuous learning, improved productivity), project teams (better quality, reduced cost of delivery), commercial teams (reduced time to create proposals, better win ratio).
Tata Chemicals	CoPs for safety, sustainability, maintenance, plant asset care, logistics, supply chain.	3,000 ideas per annum. Resilience during COVID-19 crisis for monitoring people, operations, cost. Realignment to manufacture sanitizers, chemicals for face masks.
Petroleum Development Oman	Forums for collaboration across staff and contractors.	Sharing of insights, practices and lessons during COVID-19 crisis; increase in virtual collaboration in CoPs. Millions of dollars saved in cost avoidance.
BINUS University	CoPs structured around nine Research Interest Groups (RIG), annual RIG expo.	Increased output of research papers, improved quality of teaching, move to online instruction during COVID-19 crisis.
Mobarakeh Steel Company (MSC), Iran	CoPs for "Direct Reduction" with members from MSC and group companies; inclusion of subsidiaries in future.	Increased income, reduced water and energy consumption. During COVID-19 crisis, sharing of experiences and practices for safety and WFH.
Afcons Infrastructure	Internal and external community networks for harnessing knowledge from local, global sources.	Completion of projects ahead of schedule (e.g. metro tunnels under Hooghly River), industry best practices for constructing walls.
EY	Peer-to-peer virtual networks for professional service excellence and partner engagement.	Client engagement through innovation management, hackathons, public challenges. Resilience and insights during COVID-19 crisis.

CoPs strengthen strategic partnerships with close to 50 organizations like Microsoft, Amazon and IBM. Cognizant teams also participate in the internal forums of these partners. Crowdsourcing of ideas is a common practice in such communities, and hackathons are widely used to develop solutions.

Customer CoPs are for connecting with customers to share success stories, learn the latest business and technology trends, and discover transformative digital applications. KM impacts have been realized at three levels: individuals (continuous learning, improved productivity), project teams (better quality and reduced cost of delivery via automation and reusable code) and commercial teams (reduced time to create proposals, better win ratio). The dedicated teams for KM and Innovation come under the larger umbrella of delivery excellence, and have different processes, platforms, metrics and cultural levers. For every project or engagement, specific goals are given for KM (leverage existing knowledge) and Innovation (create new technology or processes). Beyond the delivery team, there is a dedicated function focusing on NexGen breakthrough innovations.

2. *EY*

EY is a multi-disciplinary professional services organization of member firms, operating in 150 countries. It provides services in regards to tax, assurance, strategy, transaction and consulting. Peer-to-peer virtual networks for knowledge sharing and innovation are a major contributor to its success. Employees pride themselves on their knowledge, skills and willingness to help others solve problems through initiatives like EY Badges. It includes web-based, classroom learning, applied experiences and documented contributions, with a focus on future-oriented skills like data science, data visualization and AI.

EY facilitates external connections with customers, potential customers and talent, subject matter professionals, corporations and regulators in a number of ways. This includes GigNow (platform to tap into the gig economy), EYQ (EY think tank), public challenges (EY NextWave Data Science Program, EY Hackathons) and CogniStreamer (next-generation innovation management platform).

Online communities and KM tools played a key role in disseminating relevant knowledge during the COVID-19 crisis, across EY and to its partners. This was achieved through content governance mechanisms, channels for disseminating insights, visibility into data analytics and realtime news updates. Dedicated pages on the Discover portal helped clients through the volatility and uncertainty through authoritative analysis and compelling stories. Strong KM systems and communities have been critical to helping navigate the COVID-19 crisis and stay resilient.

3. *Tata chemicals*

Tata Chemicals is one of the largest chemical companies in India with operations in Europe, US and Africa. It leverages CoPs extensively, for topics like safety, sustainability, maintenance, plant asset care, business excellence, Lean, Six Sigma, logistics and supply chain. These CoPs invite industry expert views as well. Senior leadership of the company serves on multiple industry bodies such as CII, and industry forums and R&D institutes. There is extensive collaboration with leading institutions in India and overseas such as the CSIR labs, Yale University and IIT Bombay — Monash Research Academy.

There are multiple initiatives and metrics for KM and innovation, such as percentage of ideas implemented, percentage of innovations, KM Stories, KM Day, K-Fair, InnCoTech (Innovation-Collaboration-Technology), number of best practices and benchmark practices replicated. The AIM (All Ideas Matter) program generates 3,000 ideas per annum. Projects in R&D are categorized into Core, Adjacent and Transformational to allow for the influx of tacit and explicit knowledge from within the company and from collaborations.

During the recent COVID-19 crisis, the company used its domain knowledge of chemistry to re-align the factory to manufacture disinfectants (sodium hypochlorite) and hand sanitizers. It supplied more than 1.4 million litres of disinfectants and over 100,000 litres of hand sanitizers for free, to various government agencies. Materials like nano-zinc oxide — which has anti-microbial, anti-viral and anti-fungal properties — were used to manufacture face masks.

Knowledge sharing has become even more crucial during the COVID-19 crisis, for putting into place SOPs with best in class practices for monitoring people, operations and cost. Every Saturday, 'Stories of Resilience' from a company unit are shared with the Global Leadership Team. This allows an insight into how each business across the globe is shaping and re-modelling itself.

4. *Petroleum Development Oman*

Petroleum Development Oman (PDO) is the leading petroleum exploration and production company in the Sultanate of Oman. Its CoPs enhance collaboration across staff and contractors, and leverage existing tools built on SharePoint. During the COVID-19 crisis, the KM team has captured and shared lessons learned, best practices and insights around business and operational efficiencies. Interest in CoPs has accelerated, with the need for more remote virtual collaboration and networking between subject matter experts and content owners.

Significant KM benefits have been realized via millions of dollars saved in cost avoidance and cost savings — real tangible benefits demonstrated in business effectiveness. KM tools like the learning knowledge base (LKB) for sharing and applying lessons have helped significantly in this regard.

5. *BINUS University (Indonesia)*

Bina Nusantara (BINUS) University is ranked as Indonesia's top private university, with over 30,000 students, active industry links with approximately 5,000 companies, and partnerships with more than 190 universities around the world. It has won the Global MIKE Award two times, the Asian MAKE award three times, and Indonesian MAKE award 11 times. The Research and Technology Transfer Office (RTTO) was established in 2015. It governs nine Research Interest Groups (RIG) in domains like bioinformatics, photonics, food biotechnology, open source ERP, embedded systems, education technology and IoT. Each RIG is categorized as a CoP.

Intense sharing of knowledge takes place in these groups, resulting in high-impact research and scientific publications in national and international platforms. Every year, an RIG Expo is held to increase the intensity of knowledge sharing across RIGs. This includes theory as well as real experiences found in the research activities. Research collaboration is carried out both inside and outside the country with those who support Quadruple Helix Innovation (industry, government, academia, civil society).

In 2011, BINUS University, together with 20 other universities, announced the formation of the Nationwide University Network in Indonesia (NUNI). It has a work program for the development of student mobility, lecturer mobility and collaborative research programs to develop a knowledge-sharing environment.

6. *Mobarakeh Steel Company (MSC), Iran*

MSC is regarded as the largest steel maker of the MENA (Middle East and Northern Africa) region, and one of the largest industrial complexes operating in Iran. Incremental and radical innovation are managed via two different approaches in MSC. A big leap or radical innovation is managed in the R&D department, and the Excellence Department is responsible for incremental improvements. MSC has CoPs such as "Direct Reduction" that have members from other MSC Group companies. Practitioners from subsidiary companies like HOSCO, Sefid Dasht and SABA give inputs to these CoPs. Crowdsourcing has also helped in the open innovation approach.

KM impacts include increased income from producing special grades of products, reduced water consumption, reduced energy consumption, increased market value and production increase. Knowledge-sharing sessions involve local and international suppliers and vendors. MSC reports 5,202 billion IRR (Iranian real) saving from implementing 63 improvement projects.

During the COVID-19 crisis, MSC enabled knowledge transfer to share experiences, support remote workers and answer employee queries about the crisis. Top management assisted in digital transformation and online participatory tools for the collection, processing, classification and sharing of MSC's COVID-19 prevention experiences across the company and MSC Group.

7. *Afcons Infrastructure*

Afcons is one of India's fastest growing infrastructure and construction companies. It has community networks for harnessing the latest knowledge from local and global sources. They work cohesively with multiple entities such as research institutes, standards bodies, industry bodies and suppliers. Activities also extend to joint research and publication of papers. Afcons has a trademarked and copyrighted innovation framework called Improvation™. It assesses the potential of innovations based on engineering principles, customer value, differentiation and environmental standards.

The company cultivates big leaps as well as incremental improvements. One of its big leap innovations, the "Retaining Wall with Reinforced Earth Wall" has become standard industry practice. KM has also improved productivity and project management. For example, India's first twin underground metro tunnel under the Hooghly River was completed 100 days before schedule, thanks to effective knowledge transfer.

IV. In-depth Case Studies

The seven diverse organizations profiled in this section were interviewed specifically for this book chapter (see key highlights in Table 2). They represent sectors ranging from retail and digital services to automotive and real estate; one of them is a non-profit organization as well. All are headquartered in Bangalore or are the regional bases of global organizations. These seven organizations are: Bosch, Target, Trianz, Brigade Group, NetApp, Unisys and The Indus Entrepreneurs (TiE). The material on TiE's CoPs for resilience and innovation also builds on interviews conducted with TiE chapters in 14 cities in the US, Europe, Australia and Asia.

1. *Case study: Trianz*

Technology-led innovation at IT services and management consulting firm Trianz is spearheaded by the Engineering Shared Services Group. There are five large CoPs for Cloud, Analytics, Digital Studio, Business Applications, Infrastructure and Testing. They are run by the Knowledge Management (KM) team with 150–550 members in each, and 5–10 members as part of the core team within the community. Each practice has specific themes likes AI, ML, Dockers, Automations, RPA, Migrations, AWS, Azure and IoT.

Trianz has a DNA of quick adoption of new trends and technologies, which helps in having a culture of innovation and also keeps up the interest in employees. Key principles driving the CoPs are IP creation, reusable components development, products and offerings, and cross-skilling, with innovation as the core culture. All CoPs are internal, but some specific solutions are co-developed with external partners as part of community activities.

Table 2: CoPs at case study organizations.

Organization	CoP highlights	Overall impacts of KM and innovation
Trianz	Five CoPs (Cloud, Analytics, Digital Studio, Applications, Infrastructure, Testing), with 150–550 members in each (internal and external).	Proofs of Concept, prototypes, presentations, joint go-to-market approaches with partners; new market offerings. Virtual CoPs in pandemic era.
Unisys	11-year external facing Cloud 20/20™ program. Core principles: novelty, value, IP. Forums for patent filing.	Research papers, tech talks, patents, new products. Improved patent productivity. Online Demo Day during pandemic.
Target	Forums for retail-tech (data science, AI, ML); accelerator program to engage with startup community, mentorship.	Novel products from 46 startups in 7 cohorts (e.g. contactless shopping). Pitches and Demo Day 2020 held online during pandemic.
Bosch	CoPs with vertical and cross-cutting scope (e.g. digital transformation), external engagement with startups.	Expertise in automotive software architecture. Alignment and acquisition of startups. WFH tools and safe-distancing apps during pandemic.
NetApp	Internal and external forums for storage technologies, business applications. Intrapreneur and startup accelerator communities.	Patents, products, research publications. Intrapreneur Day and Startup Demo Day held online during pandemic.
Brigade Group	Forums for digital applications in real estate industry. Startup accelerator; communities with external founders and alumni.	35 startups graduated, with innovative applications in Brigade and broader industry; some raised external funds. Demo Day held online during pandemic.
The Indus Entrepreneurs (TiE)	Voluntary CoPs in 61 TiE chapters in 14 countries. Focus on startup lifecycle, business matchmaking, investments.	Launch and scaling of member startups. Mapping needs during pandemic crisis, organizing forums to build resilience.

External partners also periodically take knowledge sharing sessions for the community and provide support for competency development. Such external members include product and platform providers and startup ecosystem players, e.g. NetApp, Looker, Tableau, Talend, BluePrism, Salesforce, Microsoft, Amazon, Google, IBM, ServiceNow and Nutanix.

As it leads to creation of new offerings, both leadership and customer facing teams provide ample support to the CoPs. The CoPs interact frequently in both physical and virtual forms. Tools used include Office 365 (Teams, Skype, SharePoint) and Gitlab. Inside and outside the company, the members meet for formal and informal gatherings, which include coffee sessions and virtual knowledge shares.

Events include the 'Know Your Solution' series to showcase solutions in detail, and the monthly tech series. A particularly successful event is the Designathon, which is cross-CoP and cross-functional. Each team is composed of a mix of members of all types coming from different groups (including corporate functions) by design. A pool of senior technology mentors are made available to all teams for any high-level consulting when moving from ideation (using design thinking) to prototype or PoC stage.

As a result of the COVID-19 impact, many CoPs are conducting virtual sessions and virtual collaborations. Upcoming trends which will impact CoPs are WFH, multiple stakeholder support and project engagement pressure on key contributors. Some of the observed challenges for CoPs are continuity, budget and priorities of other tasks. These can be overcome by engaging teams, conducting regular sessions, cross collaboration across the teams, and sticking to the business needs and RoI gains.

For other companies, the Trianz knowledge and innovation team recommends keeping up regular engagement within the CoPs with knowledge sharing sessions, cross-community connects, common platforms and a rewards and recognition schemes.

2. *Case study: Unisys*

Unisys is a global digital solutions firm with large corporations and government agencies as its clients, tracing its roots to the merger of mainframe corporations Sperry and Burroughs. Today, it practices innovation through a range of activities like global innovation challenges, hackathons and technical paper contests. It has a major hub in Bangalore, India, for knowledge management, technical and innovation support. The Global Innovation Practice group includes a Global Innovation Director, research heads, experts in tools and processes, and hub leads.

There is an annual external facing program called Cloud 20/20™, which has been running for eleven years as an industry–academia series of events. It connects with students to participate in designing innovative solutions with mentorship support. Shortlisted ideas are presented at a Tech Confluence event, with rewards for winning ideas. There is senior leadership support, along with a dedicated program manager and committee, consisting of Unisys India MD, VP, senior directors and others from the innovation fraternity.

Innovation activities are focused on the vertical front (e.g. travel, transport, life sciences, finance) and on emerging technologies front (e.g. microservices, cloud, IoT, security). The Cloud 20/20™ initiative addresses themes common to both, and focuses on problem statements as provided by Unisys business leaders, its clients and partners. The core values of the CoPs are novelty, alignment to Unisys and emphasis on IP. Some CoPs are internal facing (e.g. verticals, emerging technologies) while academic initiatives and Cloud 20/20™ are external facing. Unisys clients and partners, as well as analysts, academics and industry bodies like India's NASSCOM, constitute some of the external communities Unisys works with to drive innovation.

Leadership support for these CoPs is via funding, visibility, sponsorship of proofs of concept, branding, and hiring. Innovations are showcased in a Virtual Innovation Gallery, and innovation activities are profiled in the internal Innovation Portal under different buckets like campaigns and research. Knowledge assets range from research papers to code repositories. The filing of patents is evangelized and supported, and prolific inventors are roped in as "sniffers" to identify potential innovative ideas. This has helped increase "patent productivity" of employees.

Due to the coronavirus pandemic, the physical activities of many of these CoPs have now moved online. Virtual tools like OneNote, SharePoint and OneDrive are used to capture the outputs and ideas emerging from these CoP discussions.

The communities also conduct workshops on topics like design and creative thinking, make-it-yourself (MIY using Openspace), brainstorming and brain writing sessions. For example, brainstorming sessions are held weekly, while Technical Talks and demos are held twice a month.

There is a Global Monthly Innovation connect call among Global Innovation Leaders, Site leaders and Site Idea Coaches.

Some of the outputs of the CoP activities include ideas, proofs of concept, research papers, whitepapers, technical talks (internal and external) and partnerships. Outcomes include minimum viable products (MVPs), product roadmaps, patents and even new products. For Cloud 20/20's Technical Project Contest in 2020, 3,250 registrations and 319 project submissions were received from 130 colleges across India. The eleventh edition of the contest was held virtually in 2020 due to the COVID-19 crisis, with an online jury from NASSCOM, Unisys and its clients and partners. Winning ideas will later be incubated in Unisys.

The pandemic's impact on WFH has resulted in virtual sessions of the global Innovation Campaign across all business verticals, virtual brainstorming sessions and virtual idea generation workshops over Zoom breakout areas. These sessions were supported with online tech talks from SMEs and business leaders. This was followed by the virtual Hackathon which resulted in over 400 quality ideas across geographies.

The innovation heads recommend that innovation should be ingrained in the culture of the organization itself. Associates at all levels should be encouraged to think out of the box and explore different ways to do things. Associates should also be provided platforms or forums to share innovative ideas and suggestions and get engagement and support of senior leadership. Organizations should also not shut their doors to innovative ideas from outside, but nurture channels and communities through which these ideas can be sourced and taken forward as appropriate.

3. *Case study: Target*

Minneapolis-based Target Corporation has more than 1,850 stores in the US, and also has an e-commerce presence at Target.com. The India office of Target in Bangalore serves as the extended headquarters to Target and is a fully integrated part of the global team for activities like startup engagement through an accelerator program.

At Target, innovation is a part of the strategy to build a culture where members excel in the workplace. Innovation and inspiration are a key focus of its customer engagement. There are internal programs such as the CodeRED Hackathon, 50 Days of Learning, Team Member Incubator, quarterly Demo Days and the Guest eXperience Center (GXC).

Target in India also leads the company's efforts around artificial intelligence, machine learning, in-house engineering and data science strategies. Team members' professional and personal skills are nurtured through the "70-20-10" learning philosophy (70% on-the-job learning; 20% from key relationships with mentors, leaders and peers; 10% through formal learning programs). Team members are empowered to be curious and love to learn.

For example, employees in Technology and Data Science domains dedicate one day a week to learning new skills and staying apprised of the latest innovations in their field. The initiative, 50 Days of Learning (spread across the weeks of a year), encourages employees to apply their solutions to solving a business problem or creating a breakthrough solution in retail.

Solutions are showcased through Demo Days that are attended by the CEO and global leadership team. Over 500 ideas have been showcased at Demo Days since inception, many of which are readied for in-store pilots before being deployed enterprise-wide. Due to the pandemic, the Demo Days moved online in 2020, including talks and online breakout rooms for the US and India offices.

Due to the COVID-19 crisis, the Team Member Incubator program in 2020 focused on optimizing safety best practices while providing exceptional guest experiences. Ideas have been selected from the Hackathon as well as from "Innovation Hours" that have been conducted by the Target Accelerator Program (TAP) among the India team after WFH commenced.

Target in India's CodeRED Hackathon features employees from different functions who collaborate to demo new, tech-based solutions in just 48 hours. Two such hackathons are held each year, and more than 1,500 team members have participated in the past three years. Due to the pandemic, the event was held online in 2020.

Often, ideas germinate over informal chats among team members while discussing challenges or *if only* scenarios. Sometimes, they result from conversations on tools like Slack or Yammer. At other times, they are the result of intentional responses to defined challenges, or simply anchored in the desire to see incremental or disruptive ideas come alive. Several ideas that have emerged from these platforms have been piloted and deployed at Target stores and distribution centres, resulting in greater efficiencies, cost savings or higher revenues.

The Target Accelerator Program (TAP) team in India leads the corporate accelerator program (started in 2014), as well as the

Team Member Incubator. It connects the passion, vision and agility of entrepreneurs with Target's network, contacts and infrastructure. Based on clear problem statements, startups receive strategic guidance, access to Target's technical and subject matter expertise, and real-world data to work with. At the end of the 16-week program, startups present their business case to Target leaders, investors and others during a Demo Day. TAP 2020 also moved online due to the pandemic. By the end of 2020, eight cohorts will have completed. In the first seven cohorts, 46 startups participated.

The Target Accelerator Program Advisory Board consisting of external innovation ecosystem leaders and internal leaders. Target also has memberships in industry associations like NASSCOM and Retail Association of India, as well as entrepreneurship promotion organizations like TiE and and T-Hub.

4. *Case study: Bosch*

Bosch (or Robert Bosch) is a German manufacturing company of automotive parts, household appliances, and other products. Bosch India hosts one of the company's largest R&D centres outside Germany, in Bangalore. It also specializes in IT services for mobility and energy, and hosts the startup accelerator program called DNA (Discover, Nurture, Align).

Bosch has several vertical segments, each of which has its own innovation strategy. The Bosch Innovation Framework was rolled out in 2019 to harmonize the flow, but was not a new structure. For example, it encouraged innovation throughout the phase of product engineering, and not only at the entry or exit stage.

At Bosch, everyone is encouraged to be an innovator, and innovation is discussed in a range of CoPs on technology and business topics. The smaller CoPs have 15–20 members, while the larger ones include 100 to 500 people. There are CoPs on multiple verticals (e.g. Internet of Things, or IoT). Many products connect across multiple categories as well (e.g. sparkplugs, refrigeration units). A key focus in the digital age is to enable data analytics across products, and this is reflected in the CoP discussions.

The subsidiary Robert Bosch Engineering and Business Solutions (RBEI) has an accelerator initiative to engage with startups, called DNA (Discover, Nurture, Align). Startups which are discovered by the business development members are connected to the respective business heads and domain experts. Connecting with startups helps bring more entrepreneurial thinking into the wider organization. Having an entrepreneurial flavour is seen as important for those who want to become leaders in the company. External members are included in CoPs if they are open to co-creation for innovative projects with Bosch.

A central leadership team updates a set of technology topics that are relevant for Bosch, e.g. blockchain. CoPs in such domains are sponsored by the business heads if the technology solutions have leverage. There are physical spaces to showcase innovations of Bosch in the German and Indian offices. For example, there is an Innovation Gallery in Bosch India and a makerspace in Germany, which helps inspire employees.

Updates on monthly meetings and discussions are posted. Outputs of CoP activities include ideas, proposals, prototypes and partnerships. An annual Bosch Innovation Award is presented to outstanding creators, who are invited to present their ideas at headquarters in Germany to assess potential impact. Awards are given in the original house of Robert Bosch, which has been converted into a museum.

Many CoP interactions used to happen physically or in-person. Now, due to the pandemic crisis, many of them have moved online. This builds on earlier platforms like the Bosch Expert Organization, which was launched as an Intranet in 1998. Virtual collaboration tools like Bosch Connect have now been built on top, using IBM Connections. Online CoPs also have informal activities like virtual coffee sessions, where discussions happen without pre-specified agendas.

During the COVID-19 crisis, many CoPs and members identified challenges such as availability of affordable masks, particularly for healthcare workers in India. Bosch considers itself to be a highly, socially responsible company. Under the theme "Be the barrier, not the carrier," employees worked with the medical community to develop a range of components and products like masks, ventilators and sensors for indoor positioning.

Overall, the company considers itself to be "results-oriented" rather than "presence-oriented," hence virtual working is readily absorbed into the norm. However, some activities like "walk and observe" and serendipitous interactions are happening much less now. The value of trust has become more important in a world of virtual work. Open flows can also lead to information overload, which calls for working groups to address how to reduce such excess.

5. *Case study: NetApp*

NetApp is a cloud data services and data management company headquartered in California, with a major office and startup accelerator program in Bangalore. Established in 2004 as part of the office of the CTO, the NetApp Advanced Technology Group (ATG) focuses on storage industry technologies and their business applications. The company also takes an outside-in view of innovation, and engages with innovators in the external ecosystem through its NetApp Excellerator program. Working with Indian startups, it explores emerging markets, new business models and technologies of the cloud era.

The group does longer range research in India with 15 universities as well, such as the Indian Institutes of Technology (IIT). It offers fellowship programs for collaboration between theoretical computer scientists from Indian universities and innovative engineers at NetApp. There are also partnerships with industry consortia and conferences like Storage Networking Industry Association (SNIA) and High Performance Computing (HiPC). This enables NetApp employees to collaborate and present research papers, thereby fostering innovation.

The innovation initiatives include hackathons to generate ideas that can be productized, thus building an innovation pipeline while engaging engineers. An Intrapreneurial Chapter was launched to promote innovation and leadership at all levels, from interns and new hires to the executive level. The Global Centre of Excellence provides world class amenities and an environment for innovators to drive customer success. The Data Visionary Engineering Center (DVEC) provides interactive, high-touch experience to its customers and partners in Asia, in the area of digital transformation.

NetApp Excellerator is the accelerator program to co-create with divergent thinking startups and internal domain experts. In three years of operation, it has mentored 35 startups. Collaboration is through virtual means as well as a co-working space. The selected startups are given an equity-free grant of US$15,000 each, and external investors are also invited as relevant for further rounds of funding. Startups are chosen from domains like cloud, IoT, big data and analytics, machine learning, virtualization, data security, data management, storage and other adjacent areas.

Core principles of these collaborative CoPs are trust, integrity, synergy, teamwork, simplicity and adaptability. The leadership style is democratic, collegiate and collaborative, with the aim of empowerment.

Physical activities range from the formal to informal, such as lunch discussions, meetups and cake-cutting ceremonies on achieving milestones. Higher-profile activities include the hackathons, NetApp Intrapreneur Day, University Day and NetApp Excellerator Demo Days. Awards are given for those who file patents, winners of Hackathons, Techtalk speakers, Spot awards, publishers of papers, CTO Awards and even intern showcase awards. On the Demo Day, NetApp Excellerator gives awards to startups in its accelerator in categories like Most Innovative Product, Best Growth Strategy and Investors Choice.

Outputs of the innovation activities include ideas and prototypes as output for the internal innovation programs, and partnerships that arise out of the NetApp Excellerator program. Selected hackathon ideas are included in the product roadmap. Close to 90 patents were filed from NetApp's India office in the last one year, and 70 patents granted.

Despite the COVID-19 crisis and the series of lockdowns, none of the above innovation activities stopped; they pivoted from physical space to the online mode. All the communications are virtual via platforms like Zoom and Microsoft Teams. For example, the NetApp Intrapreneur Day was held online in June 2020. The sixth cohort of NetApp Excellerator was organized completely online.

Looking ahead, some of the typical challenges faced in the CoPs is initial resistance to change. But the strength of the innovation DNA leads to eventual overcoming of this resistance. As a recommendation to other companies, NetApp recommends that innovation be part of the Key Result Area of each engineer, which is reflected in the high number of patents that are filed at NetApp India since inception. Innovation must be organic and part of the company culture, with clear measures and metrics. Recognition and rewards also help communities maintain the culture of innovation.

6. *Case study: Brigade Group*

The Brigade Group is a leading player in the real estate space in south India, ranging from property development and management to hospitality and education. It spans commercial and residential properties, and one of its notable CSR projects is the Indian Music Experience, India's first interactive museum dedicated to the country's traditions of classical, folk and contemporary music. To accelerate the pace of digitization in its operations and properties, the group has launched the Real Estate

Accelerator Program (REAP) to engage with startups; some of the startups have worked with other industry players as well, and have raised funds from external investors.

So far, 35 startups have graduated, and there are plans to raise the total to 100 within the next five years. Startups are chosen based on their ability to solve problems, harness emerging trends, and potential to raise funds from external sources. For example, emerging trends due to the COVID-19 pandemic are contactless technology in hotels, mobile check-ins, UV cleansing for phones and cabinets, and HVAC filtration technologies that can kill microbes. WFH has also led to demand for support in process automation, e.g. AI for legal documentation, registration for deeds.

REAP is regarded as Asia's first real estate tech-focused accelerator, and probably the second in the world, according to company sources. REAP adds to the R&D and innovation strengths of the Brigade Group thanks to its connects to external communities of entrepreneurs. Benefits accrue to not just Brigade, but the real estate industry as a whole. The startups work on live projects in the real estate space and 40% of them have raised additional funds. The real estate sector is generally conservative and regarded as behind the curve with respect to digital transformation; the startup community engagement helps Brigade build these new strengths.

Key focus areas at REAP are data analytics, AR/VR, robotics, nanotech, construction technology, supply chain management and environmental sustainability. Out of 150 applications per batch, five are selected; two batches are completed each year. Selection is determined by filters at three levels: REAP (choosing 15–20 startups from the 150), department heads and internal/external experts from companies like Cisco, Intel, Microsoft and CBRE (narrowing down to 10 startups), and the group chairman as well as heads of venture funds and PE funds (the final five). REAP's 16–18 week schedule for startups covers business plans, customer engagement, legal and accounting, and investor pitches. Founders also share stories with peers in the group. They get visibility within Brigade via townhall meetings, and externally to industry and industry conferences. The startups interact with Brigade employees through community events like knowledge series, demo days and even 'speed dating' events. Brigade also invests in some of the REAP startups.

Outcomes of the REAP initiative for innovation include impactful business solutions, industry recognition and third-party fundraising. All REAP startups received business from the real estate industry.

The Brigade Group itself gave business to 60% of the REAP startups. Overall, the startups delivered impact with respect to cost savings, revenue generation, reduction in turn-around time (TAT) and reduction in carbon footprint. The startups have received recognition and awards from the NASSCOM-CBRE Disruptech 2019 Conference, KPMG's Real Estate Innovation Overview 2019, Accenture Venture's Applied Intelligence Challenge 2019 and others.

Due to the COVID-19 pandemic, all the interaction events at REAP have moved online, via platforms like Zoom. The focus on adoption of technology, particularly digital technology, will be even more acute in the post-COVID era. Examples include technologies that are focused on contactless services and sanitization, e.g. contactless check in/out at hotels, contactless elevator movement, contactless opening/closing of doors and sanitization of handrails.

Younger generations of employees and particularly startups are tech savvy and keen to innovate using new technologies. The mandate for REAP is to keep the momentum going for such innovations, show the tangible business impacts, and scale up the startups rapidly. Online communities on platforms like WhatsApp help the startup founders discuss issues, keep up with the news and solve their concerns. Alumni founders and current startups in REAP also share successes and even failure stories at events and forums.

7. *Case study: The Indus Entrepreneurs*

To build an ecosystem for entrepreneurship in Silicon Valley for Indian-American startups and to strengthen the US-India corridor for innovation, the non-profit organization The Indus Entrepreneurs (TiE) was launched in Silicon Valley in 1992. The mandate expanded and in 1998, TiE chapters in Boston, Los Angeles and Seattle were started. In 1999, three TiE chapters were launched in India: in Delhi, Bangalore and Mumbai. There are now 61 TiE chapters in 14 countries, with over 15,000 Charter members. Entrepreneur members are Indians as well as nationals of other countries.

TiE city chapters operate through a combination of full-time staff and Charter members. The Charter members pay an annual fee for the privilege of being a part of TiE, and as a gesture of giving back to society by nurturing the next wave of entrepreneurs. Membership is purely voluntary, and Charter members are usually business professionals, successful entrepreneurs and thought leaders who are willing and able to help startups. Charter members interact through CoPs, which meet regularly in formal and informal physical settings as well as collaborate via online platforms like e-mail lists, WhatsApp groups and Zoom sessions. Informal activities include mixer sessions, festival parties and even wildlife outings.

Each chapter is run independently, but coordination is facilitated by the nodal hub called TiE Global. TYE (for high school students) was launched in 2008 and 30 chapters ran this program in 2020. TiE University (earlier called TISC, or TiE Startup Challenge) was launched in 2012 and 18 chapters ran this program in 2020. TiE Women was launched in 2019 and 26 chapters are running this program in 2020. Formal activities held by TiE Bangalore, for example, include the IoT Next Forum, Leapfrog and SMAC Day (Social, Media, Analytics and Cloud).

Most CoP activities over the years have focused on startup lifecycle phases like ideation, prototyping, product-market fit, launch, team-building, operations, scaling, fundraising and alliance strategies. During crises like the Dotcom bust, the financial crisis of 2007–2008 and the recent coronavirus pandemic, TiE CoPs have focused more on innovation resilience activities so that startups can be assisted in 'survive, revive and thrive' modes. In interviews conducted with 15 international TiE chapters during the pandemic in the period March-August 2020, the chapter members and CoP representatives shared how they regrouped to help innovators "pause, pivot or persist."

For example, TiE Mumbai spoke to founders on a regular basis during the pandemic to map out challenges faced during the months of lockdown, and design online forums to meet these needs (see Table 3). TiE Global held special sessions on business and mental resilience. TiE Ahmedabad, TiE Delhi-NCR and TiE Silicon Valley formed partnerships with governments and educational institutes to offer workshops and grants. TiE Kerala connected founders to investors. TiE Kolkata connected startups to experts in the supply chain to help overcome the disruptions from lockdowns.

TiE New York and TiE Melbourne designed community forums around locally popular cultural formats like open mike sessions and coffee meetups, respectively. TiE Bangalore conducted its school education program on entrepreneurship entirely online, valuing entrepreneurial thinking as key for young minds during the uncertainty of a crisis. TiE Pune

Table 3: How TiE Mumbai mapped resilience sessions for startups during the pandemic in 2020.

Month	Problems	Solutions
March	Panic, confusion, chaos, funding freeze, lack of information, no experience of business cycle disruption.	Sessions on empathy, inspiration, health, WFH, legal issues; peer buddy network; connections with legal, delivery; mantra — "cash is king".
April	Cash management, customer payments, payroll; digitalization, supply chain breaks, loss of retail; pivots; rumors, no sign of lift up, what's the "new normal".	Mentorship, idea validations for pivots, investor sessions, story sharing, Founder's Forum.
May	Transformation, turnaround, shutdowns, freezes; expense management, furloughs, cuts; "painkiller versus vitamins", moment of truth.	Masterclasses, re-skilling, mentoring; applied knowledge, problem solving; sessions with investors and angels; more SIGs, all-Chapter collaboration.
June	Organizational restructuring after pivots; renewed fundraising, conversations on valuation; acceptance of new realities setting in, scenarios of post-Covid world.	Academy courses, masterclasses on digitalization, e-commerce, HR; live demos; "startups as the new MSMEs".
July	Uneven ending of lockdowns and opening of markets; funding; ban on Chinese apps like TikTok (problem, opportunities)	Sessions and classes on marketing, e-commerce, specific sectors; sharing of first-hand experience.

arranged confidential one-on-one sessions between founders and mentors. TiE Hyderabad will be hosting the largest global TiECon in December 2020, entirely online.

All chapters conducted online sessions spearheaded by Charter member CoPs, in formats like webinars, pitch sessions, mentorship clinics, workshops, masterclasses, founder forums and demo days. Lessons were shared by senior leaders who had weathered crises from the past and shared hard-earned insights. Though the pandemic has dealt a severe blow to sectors like travel and live entertainment, it has opened up new opportunities in healthcare, education, online entertainment and virtual workflow management. TiE CoPs helped entrepreneurs form partnerships to explore new business avenues, though the challenge was how to choose a pivot that could be sustainable and viable even in the post-COVID era.

V. Analysis

As seen from the diverse case profiles, many organizations are in varying stages in the maturity curve for harnessing CoPs. Some refer to CoPs as forums, groups or networks, but are not fully capitalizing on the abundant literature and case studies for CoP definition, creation and optimization. Others have clearly-defined CoPs for internal activities, usually productivity and innovation. More mature organizations have expanded CoPs to cross-sectoral and cross-organizational knowledge activities involving business partners and customers.

The most resilient organizations have quickly adapted to the changed reality of the pandemic, and increased membership diversity in communities, quality of interaction and focus on robustness. This has helped them figure out whether to 'pause, pivot or persist' in the face of new challenges and opportunities. The pandemic has stepped up the pace and demand for resilience, entrepreneurial thinking and co-creation, all of which can be assisted by effective communities.

Practices for knowledge management and innovation are intersecting as corporate communities discover new ways of engagement with start-ups, such as accelerators, and begin to improve the performance of this model. Industry awards for knowledge management and innovation have traditionally been separate, but new research groups like the MIKE study group for the Most Innovative Knowledge Enterprise awards are blending the two.

Analysis of the case studies also shows that impacts of CoPs are felt at multiple levels: individual members, CoP leaders, specific business functions, overall organizations and ultimately society at large. Practices like storytelling and narratives in CoPs need to be harnessed even more in order to bring out qualitative and quantitative factors of community dynamics. Impacts of CoPs can be broken down into the following five types of metrics: activity, process, knowledge assets, employees and business (see Table 4).

In addition to "making rich companies richer," it is also satisfying to see CoPs being applied in government, academic and civil society settings, though more research and implementation is called for in this regard. For example, the case studies of BINUS University (Indonesia)

Table 4: CoP metrics.

Scope of metrics	Sample parameters
Activity metrics	Number of CoPs, members, messages, queries, session duration, session frequency, diversity of membership, number of blogs/assets captured.
Process metrics	Faster response times to queries, tighter collaboration, more secure communications, integration with content assets, value-added services.
Knowledge metrics	Number of ideas submitted, number of knowledge assets re-used, best practices created, rate of innovation, knowledge retention.
Employee metrics	Better employee engagement, improved soft skills, peer validation, empowerment, trust, decrease in time to competency, increased motivation.
Business metrics	Reduced costs, less travel costs, greater market share, increased customer satisfaction, profitable partnerships, conversion of knowledge assets into patents/licenses, improved productivity, risk reduction, crisis management.

and the TiE chapters show that communities of innovation can benefit educators and social entrepreneurs as well. Other groups like KM4Dev (knowledge management for development) have highlighted more such examples.

Trends to watch in the field are the rise of digital technologies for understanding and predicting conversational moves, such as AI and ML used for analytics and chatbots. Within organizational settings, social network analysis (SNA) has emerged as a powerful tool for mapping knowledge flows and identifying gaps. Natural language techniques, visualization tools and recommender systems can be harnessed in CoPs for cross-organizational and cross-functional enhancement. The use of AI and ML is becoming an important embedded layer in organizational conversations and communities in this regard.

The rise of networked virtual environments for CoPs, especially accelerated by the pandemic, throws up a number of challenges and opportunities. Combining video conferencing, chat windows, shared open documents and instant messaging can lead to new formats of knowledge assets and documents. Video can add richer dimensions to communication as compared to email. Reusability of digital content is very high, and computational techniques like AI can be used for real-time transcription and even multilingual translation. However, one of the challenges of moving largely online is missing out on the serendipitous encounters and informal exchanges during the traditional "water cooler" sessions. Communities will have to work extra hard to bring this level of openness and trust into online forums; efficiency is not the only success factor for a CoP.

In sum, despite the ups and downs of the journeys, communities of innovation have emerged as a key success factor for organizational success. Designing and launching a CoP are important steps, but maintaining, harnessing, sustaining and scaling it are equally important, as shown in the case profiles of this chapter. The coming years will continue to reveal more frameworks and frontiers for effective CoPs, which will be a most welcome move as humanity confronts even tougher challenges ahead like climate change.

Further Readings

Interviews with winners of the MIKE awards (Most Innovative Knowledge Enterprise)

1. Vineet Jain, Knowledge Leader, EY Global Delivery Services https://yourstory.com/2020/08/cii-knowledge-management-ey.
2. Richard Lobo, Head of Strategy and Business Excellence, Tata Chemicals https://yourstory.com/2020/07/tata-chemicals-knowledge-management-richard-lobo.
3. Hank Malik, Knowledge Management Program Lead, Petroleum Development Oman https://yourstory.com/2020/07/cii-petroleum-development-oman-hank-malik.
4. Elidjen, Knowledge Management and Innovation Director, BINUS University, Indonesia https://yourstory.com/2020/07/cii-knowledge-management-binus-university.
5. Akbar Golbou, Knowledge Management Team, Mobarakeh Steel Company, Iran https://yourstory.com/2020/07/cii-knowledge-management-mobarakeh-steel.
6. Hariharan Mathrubutham, Vice President for Knowledge Management, Cognizant https://yourstory.com/2020/07/intelligence-cognizant-knowledge-management-ai.
7. Rudolf D'Souza, Chief Knowledge Officer, Afcons Infrastructure https://yourstory.com/2020/07/afcons-infrastructure-knowledge-management.

Interview sources for case studies:

1. Ved Prakash, Chief Knowledge Officer, Trianz.
2. Manoj Hariharan, Chief Knowledge Officer, Bosch Engineering and Business Solutions.
3. Nirupa Shankar, Executive Director, Brigade Group.
4. Madhurima Agarwal, Director, Engineering Programs, NetApp.
5. Ravi Shankar Ivaturi, Business Operations Senior Director, Products and Platforms, Unisys.
6. Suma Ramachandran, Communications Lead, Target.

Interview sources for TiE chapter case studies:

1. TiE Bangalore: Vijetha Shastry, Executive Director https://yourstory.com/2020/07/tie-bangalore-coronavirus-covid19-entrepreneurship.
2. TiE Global: Vijay Menon, Executive Director https://yourstory.com/2020/08/tie-global-coronavirus-startups-resilience.
3. TiE Silicon Valley: Neha Mishra, Senior Director https://yourstory.com/2020/08/tie-silicon-valley-coronavirus-startups-resilience.
4. TiE Pune: Vandana Saxena Poria, Charter member https://yourstory.com/2020/07/tie-pune-startups-coronavirus-entrepreneurs.
5. TiE Kerala: Nirmal Panicker, Executive Director https://yourstory.com/2020/07/tie-kerala-startups-coronavirus.
6. TiE Hyderabad: Phani Pattamatta, Executive Director https://yourstory.com/2020/07/tie-hyderabad-startups-coronavirus-resilience.
7. TiE Chennai: Akhila Rajeshwar, Executive Director https://yourstory.com/2020/07/tie-chennai-entrepreneurs-coronavirus-resilience.
8. TiE Kolkata: Abhranila Das, Manager https://yourstory.com/2020/08/tie-kolkata-coronavirus-startups-resilience.
9. TiE Mumbai: Naveen Raju, Executive Director https://yourstory.com/2020/08/tie-mumbai-entrepreneurs-resilience-coronavirus.
10. TiE Ahmedabad: Piyalee Chattopadhyay, Executive Director https://yourstory.com/2020/08/tie-ahmedabad-entrepreneur-resilience-coronavirus.
11. TiE Delhi NCR: Geetika Dayal, Executive Director https://yourstory.com/2020/08/tie-delhi-ncr-resilience-entrepreneurship-coronavirus.
12. TiE New York: Dharti Arvind Desai, President https://yourstory.com/2020/08/tie-new-york-entrepreneurship-resilience-coronavirus.

13. TiE Melbourne: Saurabh Mishra, President https://yourstory.com/2020/08/tie-melbourne-entrepreneurship-coronavirus-resilience.
14. TYE: Geetha Ramamurthy, Founder https://yourstory.com/2020/08/tye-bangalore-tie-entrepreneurship-students https://yourstory.com/2020/08/tie-tye-bangalore-entrepreneurship-education.

References

Beerli, A., Falk, S., and Diemers, D. (2003). *Knowledge Management and Networked Environments: Leveraging Intellectual Capital in Virtual Business Communities.* AMACOM.

Cross, R. (2004). *The Hidden Power of Social Networks.* Harvard Business Review Press.

Feld, B. (2012). *Startup Communities.* John Wiley.

Gannett, A. (2018). *The Creative Curve: How to Develop the Right Idea, at the Right Time.* Currency Publications.

Hansen, S. (2017–2019). *The Startup Guide* city series.

Jansen, S. and Pieters, M. (2018). *The Seven Principles of Complete Co-Creation.* BIS Publishers.

Leonard-Barton, D., Swap, W., and Barton, G. (2014). *Critical Knowledge Transfer: Tools for Managing Your Company's Deep Smarts.* Harvard Business Review Press.

Rao, M. (2004). *Knowledge Management Tools and Techniques.* Butterworth-Heinemann.

Shelley, A. (2017). *KNOWledge SUCCESSion: Sustained Performance and Capability Growth through Strategic Knowledge Projects.* Business Expert Press.

Tan, M. and Rao, M. (2013). *Knowledge Management Initiatives in Singapore.* World Scientific Publishing.

Välikangas, L. (2010). *The Resilient Organization.* McGraw-Hill.

Part 6
Conclusion

© 2021 World Scientific Publishing Company
https://doi.org/10.1142/9789811234286_0015

Chapter 15

The Practice of Communities: How Can Society Mobilize Communities of Innovation

Benoit Sarazin, Laurent Simon and Patrick Cohendet

The cases discussed in this book show that communities are a key socio-cognitive device for innovation in and outside organizations. They allow to feed and fuel organizations with the innovative ideas of passionate members. Yet, mobilizing and harnessing communities isn't an easy task. In this chapter, we suggest a few recommendations to reap the benefits from communities. We review the hypothetical case of a firm developing its relationship with a community. We believe the principles and recommendations that we're drawing out of it would also apply to public organizations.

I. Finding the Fit Between the Community and the Organization

As discussed in Chapter 1, there are many ways to define communities: Communities of practice, communities of users, virtual communities, communities of interest, epistemic communities, internal or external… In this book, for the sake of simplicity, we suggest regrouping these categories under the name of "communities of innovation", as they all have in common that they bring a significant value for the innovation processes of organizations.

However, operational issues may be quite different whether the members are within or outside of the boundaries of the organization. For these reasons, in this chapter, we're making a distinction between three categories: internal, external and hybrid communities, where members can belong to an organization and a community across the boundaries of the organization.

1. *Internal communities*

Internal communities are basically communities of practice or communities of specialists. They gather employees of the firm who are sharing the same practice, the same occupation or are working on common topics, without necessarily belonging to the same team, department or division. They are internal to the firm and don't include external members.

We differentiate two types of internal communities:

- **Unpiloted communities:** They emerge organically without any formal input from the hierarchy of the firm. They operate in a very informal manner with no official monitoring: they don't record their members and their number is not precisely known. These are for instance the first constituted communities at Schneider Electric (Chapter 4), before the firm would launch a program to monitor and support communities.
- **Piloted communities:** They are sometimes fostered, and at least monitored by the firm, that suggests a structure and some tools to support all the communities. They officially record their members and report the outputs of their work. Innovation communities at SEB (Chapter 5) or Schmidt Group (Chapter 6) are examples of piloted communities of practice.

2. *Hybrid communities*

Hybrid communities include employees of the firm and external members. Some of them can work for the firm, with others coming from suppliers, distributors or being clients or users.

They can also come from organizations outside the sector of the firm yet belonging to its wider ecosystem. For instance, the members of Renault community (Chapter 7) don't only belong to Renault or suppliers like Valeo, they're also coming from research centres on nuclear energy, universities chairs on innovation, firms outside of the automotive industry like Air Liquide, L'Oréal or Électricité de France. Some of them belong to civil society like journalists, artists and even philosophers...

In this book, Renault innovation or Michelin Open Lab (Chapter 8) are hybrid communities. Communities of specialists at Ubisoft (Chapter 2) are also examples of a hybrid community: even if most of their members are employees of Ubisoft, these communities also include external members from outside.

3. *External communities*

External communities include members that are external to the firm yet with cognitive interests close to the activity of the firm. We differentiate two types of external communities:

- Communities of users of the products of the firm. They gather because they share a common passion which practice develops through the use of the products of the firm. These are, for instance, the trail runners wearing shoes from Salomon (Chapter 3), gamers playing *Trackmania* (Chapter 9) or communities of fans with Lego or Ankama (Chapter 12).
- Communities of specialists from occupations that are not present inside the firm. These could be for instance, the historians who took part in the development of the *Assassin's Creed* series with Ubisoft developers (Chapter 2). These also could be the podiatrists with a sport expertise who helped fine-tune the trail running shoes of Salomon (Chapter 3).

These communities bring elements of knowledge that are complementary with those of the firm to start up or accelerate innovative projects.

4. *Choosing the right type of relationship*

With external communities, several types of relationships are possible, and the firm should choose the most appropriate. These are the most common:

1. Fostering a direct favoured relationship with community leaders, acting as relays with the other members. That's the case with Salomon, sponsoring top performing athletes admired by the community of trail runners (Chapter 3) or with online users (Chapter 12). This configuration is particularly adapted when the leaders have a strong reputation with the most active members of the community.
2. Implementing crowdsourcing platform. This technique allows to potentially create a direct relationship with the widest part of the users, without the intermediation of community leaders. That's how Decathlon Création platform invites all users to contribute to product improvement and innovation (Chapter 3). This is especially appropriate when it's not possible to identify specific leaders and to reach a wider base of potential contributors form the community of users.
3. Providing the community with creation tools to support the creative contributions of the most engaged community members. That's how Lego shared knowledge with Adult Fans of Lego to give them the means to create their own models (Chapter 12). This option is relevant when part of the product can be modified and improved by users, as it is the case with some software and video games.

5. *Contributions that the firm can expect from communities*

While contributing to the openness of the firm, communities can bring the elements of knowledge necessary to answer innovation requirements, especially by aligning further with users' needs. As a matter of fact, only innovating with the ideas of R&D and Marketing departments from the product point of view presents some limitations. The challenge is to adopt a more systemic viewpoint beyond the technical features of the product: the variety of users and usages, the environmental externalities, ethical issues, symbolic dimensions and opportunities for customization, etc. By dialoguing with communities in real time, the firm benefits from a direct connection with potential contributors possessing the missing elements to the innovation equation to answer to these new issues.

Concretely, the main benefit brought by the community is to grant access to an external source of creative ideas: this acts as a creative slack developed and maintained by community members (see Chapter 1).

Indeed, driven by their passion, community members would continuously generate ideas. They have access to the experiences lived and ideas expressed by all other members.

Further outcomes can be identified, according to community types:

- Internal communities allow to:
  - — identify best practices and facilitate their sharing with other members and employees.
  - — compensate for the lack of circulation of information between services and departments by overcoming the effect of silos and organizational barriers that are naturally occurring between isolated teams.
- Hybrid communities allow to:
  - — widen perspective by connecting to external knowledge bases and experiences that are not covered by the existing competencies of the firm;
  - — generate creative ideas thanks to the dialogue with persons with different worldviews.
- External communities allow to:
  - — enrich the products of the firm. Community members can provide genuinely new ideas inspired by their experience of usage or by their applied expertise. They can create complements to products, prototype and test them with other members, as illustrated by Lego fans sharing them freely to enrich the playing experience of the other fans;
  - — accelerate the development, prototyping and fine-tuning of products, as with Lego again, Salomon or with the members of the Ankama community or Ubisoft fans, testing beta versions and giving feedback to internal development teams;
  - — generate some buzz for a product before and after its launch, expressing their enthusiasm and promoting the product for free;
  - — identifying the best talents among community members, to help the firm in its talent scouting, spooling and recruiting.

II. Assessing the Needs and Expectations of Community Members

A community draws its strength from the passion and enthusiasm of its members for their topic of interest. However, the informal nature of communities makes them fragile: community members don't have hierarchical structures to hold them together and ensure their sustainability.

In order to support the dynamics and resilience of communities, we address hereafter the main needs of a community.

1. *A strong motivation*

As passionate as they may be, members need to be motivated to engage in the activities of the community. Participation is on a volunteer basis, with no obligation. Members' participation will be fostered by three incentives:

- Practising their passion, keeping on learning and sharing their practice, be it connected to their occupation or not, and personally benefiting from progress and improvements produced by the community.
- Peer recognition stimulates higher level of engagement for certain members, beyond the material benefits they can draw out of it.
- Belonging to a group and sharing the same values.

2. *A climate of trust between members*

Trust is the essence of a community. Trust is generated and sustained when several elements are present:

- The creation of active ties between community members. Fostering and supporting interactions between members will help in building trust.
- The contribution principle: each member accepts to contribute to the common good of the community and to share what he/she knows with other members. This contribution is based on free will, without any expectations of immediate material return.
- Autonomy, free of any formal or statutory authority or domination. For the community to thrive, members must feel free from any statutory, internal or external hierarchy, or formal constraint. A community can welcome inspiring leaders but will reject imposing gurus trying to decide for the others. This also means that the community won't tolerate that an external hierarchy — like the one of the firm — would impose a direction.

3. *Members' inspiration*

Members of a community have at least a common interest. They need a shared purpose and an inspiring mission, a mobilizing cause. This purpose is generally formulated by the community leaders, translating the interest of members into one agreed upon mission.

The engagement of the community on a mission is often expressed by a manifesto, a shared declaration of intention, endorsed by all members and supported by a set of values that are acknowledged and defended. Be it tacit or explicit, the manifesto is key to the core identity of the community.

Members need a shared language, recognized methods and a common set of tools to implement the ideas or practices of the community. For instance, the rules and conventions developed by Ankama fans are key to the coordination of online competitions for the players (Chapter 12). These elements constitute the "codebook", allowing the manifesto to take a concrete shape. It describes the specific language of the community, its way of seeing and doing things, a "grammar of use", a quasi-manual of its ideas and best practices.

4. *The animation of the community*

The community needs animation and facilitation to give density and rhythm to its activities and to foster and maintain the engagement of members.

This is the role of community managers, who organize events and stimulate members to express and share their ideas and to launch projects. They also channel discussions between members, for instance by making sure that questions asked on social networks would get an answer and foster discussions. Without animation, the community is at risk of waning. As soon as they feel that activities are declining, community members lose interest to refocus elsewhere. For instance, they can disengage easily if they aren't challenged to participate, or because they feel disappointed by the lack of reactions of other members to their inputs.

Any community needs rules to support its operations. These rules are conceived inside the community and allow the members to frame their expectations and orient their relationships and behaviours. They must be aligned with the values of the community, tacit or explicitly expressed in the manifesto.

For instance, the Renault community adopted as an operating principle the suspension of critical judgement on other members: only ideas are to be debated, in order to feed discussions and enrich knowledge bases for the benefit of all members (Chapter 7).

III. The Middleground, an Essential Link Between the Community and the Firm

Let's draw on the metaphor of the gardener…

The gardener takes care of the garden by multiple actions: preparing the soil, watering the plants, feeding them with fertilizer, protect them from winter freeze or direct sunlight, and in the end selecting and harvesting them. Yet, the gardener can't pull on the leaves of the plant to make it grow!

In the same spirit, the firm can take care of the context in which the community is going to root and thrive, but it can't direct it. We coined this context as the middleground. The middleground can be defined as a support context and a set of key activities through which the community is going to grow, develop and thrive. As the gardener would nurture the soil with fertilizer, the firm should feed the community with a sense of purpose, challenging questions, elements of knowledge, and actively foster links between members and eventually with other communities.

A healthy middleground is a common playground and a set of "commons" co-created and shared by the community and the firm:

- It provides the context for the community members to share knowledge and debate, and supports the development of the community.
- It sets the stage for relationships and exchanges between the firm and the community and allows the firm to capture new inputs and ideas and convert them into innovation.

The middleground takes shape through four different yet mutually complementary dimensions:

1. *Places*

Places are physical or virtual locations for informal, unplanned gatherings.

As an example, Ubisoft designed special areas between regular workspaces to encourage informal encounters and discussions, and internal galleries for artists, with lounge furniture, that create opportunities to gather and share ideas between members of different communities (Chapter 2). The communities of trail runners using Salomon products gather informally on the country paths and chalets they're using for training (Chapter 3). Salomon even created, sponsored and animated open mountain chalets to foster and facilitate these encounters and conversations. Places can also be virtual, as for Ankama's multiple communities of players and Lego's amateur developers, sharing their views and insights through online forums, dedicated blogs or Facebook accounts (Chapter 12).

2. *Spaces*

Communities are driven by meaningful and purposeful questions and defining subject matters that orient collective questioning, debates between members and research activities. These cognitive spaces allow the members to define their playground and to develop their thinking by sharing with other members or even other communities, local or distant. Spaces appear as the chosen, significant topics for a community. For instance, rethinking the social output of transport and mobility for Renault community (Chapter 7) or for Michelin (Chapter 8), debates on the virtues and limits of virtual reality devices for Ubisoft game designers (Chapter 2), or running on the sole or on the toes for Salomon trail runners (Chapter 3)…

3. *Events*

Community members gather for events where they can live significant moments of sharing together and can also sometimes open to external stakeholders. They share knowledge, ideas, insights, and meet with new members. The serendipity of events supports the development of new ties between members and sparks new ideas through new connections. For instance, the SEB community organizes yearly Innovation Forums to bring together all its members, and regular Innovation Days on specific topics for sub-communities (Chapter 5). Ubisoft organizes weekly "Hot Fridays" when members can share about the development of their projects. These events act as special opportunities for connections between formal projects and diverse communities (Chapter 2). Trail runners get together for competitions, and the celebrations that follow are considered as privileged moments of sharing (Chapter 3).

4. *Projects*

The most involved members often launch projects through which they aim at making the ideas of the community concrete. These projects help in validating the value of the knowledge of the community. They give shape to the ideas that are emerging from the connections between members.

The hackathons, events organized by Hacking Health, aim at fostering projects with participants from different communities over a 48-hour sprint (Chapter 10). The "crisis community" also engages a diversity of stakeholders into concrete actions through the implementation of specific projects (Chapter 13).

By contributing to the middleground, the firm shows that it cares for the community. It creates a trusting relationship with the members without being perceived as invasive.

It gets internal communities to open to and connect with the pool of creative ideas of employees. For instance, Ubisoft organizes internal ideas challenges through which the firm can reveal and capture new ideas that could be useful for present and future projects (Chapter 2). With external communities, the firm opens to extended networks of knowledge and connects its employees with external stakeholders with complementary knowledge assets. Without the mechanism of the middleground, it would be almost impossible to organically connect with potentially useful new ideas of individuals and communities, be they internal or external to the firm.

Before defining its relationships and contributions to the middleground, the firm must make sure to set up the right context and conditions.

IV. Supporting the Community

1. *Understanding the manifesto and the codebook*

The manifesto and the codebook are essential drivers of the community and it is key to understand them in order to grasp what motivates the members of the community. This allows to clarify the domain that is of interest for the community and the firm, to foster fruitful interactions. It happens quite often that the manifesto and the codebook are tacit. Representatives from the firm can investigate them through the following questions:

- What is the actual passion that unites the members?
- What is the purpose and mission of the community? What kind of questions and challenges is it trying to tackle? What kind of "commons" is the community trying to build?
- What are the underlying values of this mission? How and why are they significant and meaningful for the community?
- How is the community different from what is already existing? Is the community disruptive in any way? What are its distinctive characteristics and features?
- What are the practices that are shared by the members? What makes them genuine and distinctive?

2. *Respecting the values of the community*

A community would always express strong values, implicit or explicit in its manifesto.

It is essential to respect these values, otherwise the firm is at risk of being rejected by the members.

One example of rejection is illustrated in Chapter 3 where the leaders of the trail running community pushed the track and field federation back, as they would estimate that some of its endeavours where against the values of the community.

3. *Fostering an open and transparent dialogue with the community*

The members of a community are used to an open, genuine and transparent dialogue. They wouldn't accept discourses driven and motivated only by exploitation, or even less so discourses obscured by corporate managerial lingo. To gain credibility with the communities, the firm must adopt the same language as they're using, led by truth and the trust between persons engaged on the same mission. This means that the discourse of the firm must state things as they are, without trying to embellish the truth as it is often the case with corporate communication. The firm must be humble and recognize its weaknesses and mistakes. It means a significant culture change for managers in charge of internal or external communication.

We have mentioned the general principles for all types of communities. Now we would like to consider in finer details the specific principles that would apply to internal, hybrid or external communities. With internal communities, the firm can implement the following actions that are described in the next sections: bringing hierarchical support, measuring the value of the contribution of the community based on members' appreciation, supporting the structuring of communications and activities, fostering a safe and respectful environment, recognizing the engagement in communities when evaluating employees.

4. *Ensuring the support of the hierarchy*

The firm can ensure the support of the hierarchy to the initiatives of the community and give them some visibility inside the organization. This shows that management recognizes the value of the work of the community.

For instance, Ubisoft Montréal formally appoints a "leader" for each community of specialists, with the mission to act as an animator and as a spokesperson (Chapter 2). This title doesn't define a authority status, but rather the role of a go-between in charge to represent the community to the hierarchy and to communicate its state of mind, specific needs or issues encountered by members in the achievement of their mission. The leader also monitors the activities in order to detect the ideas that may be valuable to solve the firm's innovation needs. For instance, it is also the role of the communities' sponsors at SEB, Schneider Electric or Schmidt Group (Chapters 4, 5 and 6).

5. *Measuring the value of the contribution of the community based on members' appreciation*

It is normal and legitimate for the firm to try to measure the value that the community can bring to its activities. However, this value is difficult to objectively and quantitatively assess, based on tangible results, for two reasons:

The first one is that, when a potentially innovative idea from the community reaches the firm, it is generally rough and only partly defined, and its value is difficult to seize. It needs to enter the formal innovation processes of the firm and to get refined and combined with other elements of knowledge to achieve some tangible and measurable outcomes. In this configuration, it is difficult to determine which parts of the value comes from the community and which from the parts coming from formal inputs from the organization.

The second one is that, the direction expecting specific results would actually violate the freedom of the community members to choose their own focus of interest and work orientations. However, the direction can ask the members to give their own assessment of the value brought by the community and to argue about the way they assess this contribution. It is the evaluation mode chosen at Schneider Electric for its communities of practice (Chapter 4).

6. *Helping communities to structure themselves*

If the firm decides to support the work of internal communities, it can implement actions to help them structure their meetings, knowledge management and communications. For instance, the firm can suggest tools and facilitate the use of internal platforms and communication technologies. However, this requires caution and subtlety, and it is recommended that the direction would help only on demand when the community reaches a certain level of maturity, to respect the community's autonomy. It is the posture adopted at Schneider Electric (Chapter 4) or Michelin (Chapter 8).

7. *Fostering a safe and respectful environment*

It is important for the community to operate and grow in a safe and respectful environment. It allows to avoid that the members would feel judged or criticized for their involvement and would refrain from expressing their views. It is all the more important for new members whose contributions may not be aligned with the community's routines and can bring valuable disruptive points of view and insights. It is quite often recognized that these members are sources of very original ideas as they are not "contaminated" by the dominant thinking of the group.

8. *Recognizing the engagement in communities when evaluating employees*

An ultimate step would be to grant a formal recognition to the employees who are engaging with the communities and to integrate this in the formal evaluation for bonuses and promotions.

With the external communities, the firm can implement the following actions that are detailed in the next sections: identifying the roles in the community and fostering pride.

9. *Identifying the roles in the community*

In a community, different members tend to play some specific roles to support, animate and stimulate the activities. For instance, in the video game *Trackmania* (Chapter 9), different and complementary roles appeared: race-tracks creators, online events managers, team managers and racers, inspiring the community through their successes and competing spirit. In the hack-athons of Hacking Health, complementary contributions are possible through the engagement in different roles: understanding and exposing medical needs for healthcare professionals, mobilizing the potential of IT for programmers, shaping and prototyping ideas for designers, business models development for entrepreneurs… (Chapter 10). Identifying these roles and contributions will help the firm in providing specific and adapted support.

10. *Fostering pride*

The brand image and reputation of the firm is key to inspiring the members of external communities to engage and contribute, and to express new valuable ideas that could be integrated in the firm's products or services. For instance, the strong and positive image of Ubisoft with the community of gamers helps in attracting and mobilizing gamers and it makes them proud to help and contribute to the improvement of the games of the firm (Chapter 2). It some cases, gamers, as external contributors, can even be named in the game's final credits.

11. *Setting the rules for confidentiality and property rights*

In most communities, and specifically in hybrid communities, it is key to set up the rules for confidentiality as early as possible to avoid limiting the openness in the interactions between employees of different firms and community members. Several options are possible. It can be decided that no property rights would be applied on the outputs of the community and that every member is free to exploit them without any limit. It is the policy of the Renault community (Chapter 7). It is also possible to experiment with more open property rights, like the "creative commons". Some rules of participation can also be enforced in order to exclude competitors to work in the same workshops or on the same projects. It is the policy of the Open Lab community of Michelin, for instance (Chapter 8).

V. Supporting and Giving Without Expecting Instant and Equal Returns

The firm must accept to support the community through the middleground, without expecting instant and equal return. The community is not a space of transaction, rather of cumulative collective contribution: each member contributes to the building of the common good and shares their knowledge without expecting other returns than the access to knowledge and other members. The members of the community would expect the exact same attitude from the firm, as an equal member. By respecting these expectations, the firm would be able to earn the trust of the members, into a sustainable relationship, and eventually would benefit from the accumulated knowledge, new insights and genuine ideas of the community.

When the firm would eventually exploit some elements from the community, it must make sure to recognize the contribution of the community and to give back to the community, in support for instance, or with added knowledge and ideas learned from the exploitation. Otherwise, the firm would be perceived as a free-rider and would be at risk of exclusion.

In the following sections, we describe some actions that the firm could implement in order to give back to the community through the middleground:

1. *Provide solutions for the animation of the community*

Animation is a key element of the life of a community. Members keep participating because interesting things are happening inside the community. Animators play a defining role by facilitating knowledge sharing and circulation through social networks. They organize events on physical or virtual platforms. They support the orientation and organization of projects. They generally monitor the quality of interactions and give the right tone for respectful conversations, for instance, giving a voice to newcomers, and regenerating debates with new perspectives.

The firm can easily contribute to these animation activities in several ways:

- Participating in the casting and hiring of animators and funding their positions. Yet, the animators must remain independent as full-fledged, autonomous members of the community, with no formal ties with the firm. In general, these animators are already recognized by the other members for their knowledge and contributions and for their alignment with the mission and values of the community. With internal communities, animators may be employees delegated by the firm, with personal objectives and dedicated time, as it is not realistic, or under optimal, that an employee would animate a community over and beyond their normal working schedule.
- Supporting animation through a constant culture of feedback.
  As the community would evolve, the firm can encourage the animators to collect feedback from the members to identify their needs and adapt animation and the supporting tools. For instance, at SEB, the community coordinator spends a lot of time probing the members about their needs and trying to provide solutions with tools from the firm (Chapter 5).
- The firm can contribute with online collaboration tools.

For instance, Schneider Electric opens its corporate social network to internal communities (Chapter 4). Salomon implemented a Facebook "Salomon Running" section to facilitate the conversations between the members of the trail running community (Chapter 3).

2. ***Providing material support***

Material support is a key contribution that a community can expect from a firm, as the community doesn't generate revenues and must count on the benevolent contribution of members for its material organization. This could be financial support, to organize events, for instance, or providing technological solutions, like technological platforms for virtual interactions or knowledge management software, or even manpower, by delegating an employee as an animator. In these matters, the firm must proceed with caution, with a disinterested attitude, to avoid being seen by members as trying to "buy" the community. In practice, financial contribution and material support should be complemented with other types of actions and active involvement showing genuine interest for the "quest" of the community and not only for its products.

3. *Giving members the opportunity to make their ideas concrete and to test them*

When members of the community are generating ideas, it is quite often difficult for them to translate their insights into reality and to dedicate means to build and prototype them. The firm can play an active role here by offering some production means for rapid prototyping, for instance, or for testing. The firm can even support the incubation of ideas that would allow to transform the insights in a pre-entrepreneurial form, test them with potential consumers, and compare them to existing potential competitors. Renault, for instance, lend some electric vehicles to the community to test and experiment with the behaviours of drivers (Chapter 7). Another example is with Hacking Health, organizing hackathons on the site of a healthcare institution to support the demonstration of a prototype in real-life conditions with users (Chapter 10). This later developed as a startups accelerator channelling resources form the healthcare system and venture capitalists to expand the projects of the community to entrepreneurial endeavours with a commercial focus.

4. *Rethinking work schedules and workspaces*

When the members of the community are employees of a firm, management can arrange schedules to give some opportunity to members to engage in informal meetings and community events.

For instance, at Ubisoft (Chapter 2), the participation in informal sharing moments organized every two weeks ("Hot Fridays"...) is considered as a normal part of the working load of employees.

The firm can also re-organize the workspace in order to create places for informal meetings and discussions for knowledge sharing. Again, at Ubisoft, employees have access to multiple playful and inspiring spaces (playrooms, internal art galleries and cafés) that allow for the experience of unstructured and unplanned interactions supporting social connexions, inspirations and knowledge sharing.

VI. Enacting Ideas to Actualize and Harvest the Value of the Work of the Community

The representatives of the firm can't define the orientation of the community in its exploration of new ideas. On the other hand, it can enact some of them to support their development and put them to the test. These efforts of enactment can play the role of concrete experimentations whose results can add to the knowledge of the community and help in orienting its future investigations.

In what follows, we describe some of the actions that the firm could implement in order to facilitate the enactment of the knowledge work of the community in actual projects, via the middleground:

1. *Clarifying the strategy of the firm*

The firm must make sure that its strategic orientations are clear and understandable by the community. It also must express under what kind of constraints it is operating, to clarify what kind of developments would be realistic or not. When the firm engages with the community, it discovers many ideas on the multiple topics of interest of the members. Not all these ideas are aligned with the strategic orientations and interests of the firm. Some of the ideas may be too disruptive or too advanced for the state of technology.

The representatives of the firm must clarify their interests, and also be precise about what can be realistically be done in terms of development, in order to avoid misunderstandings and false expectations form the members. As any member, the firm should express its freedom to choose the ideas it believes to be worth exploring.

2. *Challenging the community with purposeful questions*

To play an active role, the firm should regularly challenge the community with difficult questions aligned with the purpose of the community and by suggesting investigating some specific ideas. For instance, when social networks were starting to get traction with the public, Ubisoft launched an ideas competition with its employees to find ways to engage and generate revenues with Facebook. This generated hundreds of suggestions, some of which were later explored and implemented in products.

In the same spirit, Decathlon Creation crowdsourcing platform regularly challenged the community of users and aficionados to "reinvent the camping tent" or "imagine the future products for indoor soccer" (Chapter 3).

3. *Implementing recognition mechanisms*

When one or several members of a community would contribute to the design or development of a product or a service of the firm, it is important that the firm assess and recognize this contribution. For instance, when some game fans contribute to the development of a game, they may appear in the final credits, or be identified and officially thanked on social networks of forums. In the same spirit, the leaders of the projects that emerged through hackathons are recognized by Hacking Health in final events and through formal communications (Chapter 10).

4. *Enriching the community by fostering connections with other communities*

The creativity of the community is all the more fruitful when it has the possibility to interact with other communities that would feed it with new knowledge and ideas. It is mostly in the spaces of the middleground that these connections can occur. Yet, it is not very easy and natural for a community to open to others, as members tend to stick together and gravitate around their common interests. It is easier to interact with members who share the same language, the same passion and the same worldview.

One example of this difficulty is illustrated in the chapter on Hacking Health. The organization connects healthcare professionals with programmers, designers and entrepreneurs. To overcome the difficulties of communication between these knowledge groups with different backgrounds and worldviews, Hacking Health also invites patients who help in translating different worldviews in a common purpose focused on more efficient processes for treatment (Chapter 10).

The challenge of finding a common ground for different knowledge groups actually fits the capability of a firm to bring together people with different backgrounds on a joint project. This is where the firm can help communities to learn how to interact with each other, and to open up to different worldviews and ideas.

This can bring significant results, as shown by the example of Decathlon, with the Sense Shoe project, bringing together four different communities (Chapter 3). It is also the case with Renault, inviting a philosopher to enrich the discussions in a community mostly composed of engineers (Chapter 7) or with Michelin, getting experts on robotics and exoskeletons on board to debate with mobility experts (Chapter 8).

VII. Managing with Communities

In the next sections, we address what the attitude of a manager should be to work more efficiently with communities. We identify seven components of this attitude.

1. *Strategic awareness*

To be able to mobilize communities efficiently, the employee of the firm interacting with them must first identify what are the key issues and challenges for the firm and its main customers. To do so, this employee must get a clear vision of the position of the firm in its environment: present and future needs, positioning of competitors, and assess market, economic, technological and sociological trends. This would allow to identify faster the communities that are engaged on domains of interest for the firm.

Being able to clearly express the orientations and interests of the firm also helps in managing the relationships and expectations of community members. It justifies the choices the firm could make at some point in supporting or not some orientations of the community and avoids misunderstandings and possible frustrations.

2. *Dynamic and systematic monitoring of the landscape of communities*

Communities are self-centred and generally don't communicate openly about their existence and activities. To identify the communities that could be relevant to the firm, the manager should engage in active exploratory work by mobilizing its personal network of clients, suppliers and partners; and monitoring virtual and social networks.

3. *Active observation*

When a community is identified and seems to potentially fit the innovation interests of the firm, an observation phase is necessary before actively engaging with interactions. For locally accessible communities, this could mean passive participation in events, for instance. For virtual communities, this could mean, the first time, observing online conversations on specific topics, emerging debates and trying to identify specific leaders and to seize the key interests and values of the community, in order to introduce oneself properly the second time.

4. *Candid introduction*

When ready, the manager should introduce her/himself with humility, as an amateur open to learning, and not as an expert imposing her/his views, by raising questions on the main topics of the community, on its present investigations, and on its rules and processes of interactions.

If the question is relevant and genuinely interesting for the members, they would be curious about the newcomer, and ultimately open up and engage in a fruitful dialogue. The introduction is a very sensitive moment when the manager must demonstrate her/his knowledge and ability to point to relevant challenges for the community, otherwise she/he may remain at the periphery of the community and connections with the most interesting and knowledgeable members may become an issue.

In order to build trust, the manager must also be transparent about her/his position in the firm and explain how the firm could be interested in learning but also in contributing to the community. If she/he appears as a "double agent" looking only to absorb knowledge and exploit interesting ideas, without any returns, the doors would quickly close and the manager would be at risk of being rejected and expelled from the community.

5. *Genuine feedback*

Honesty, transparency and humility appear as necessary virtues in the relationships between the manager and the community. If the manager would bring back some knowledge and ideas of the community in the firms' activities and innovation projects, she/he should also inform back the community about the results and developments, be they successes or failures. Such feedback would keep the conversation going with the members of the community as they would contribute to the advancement of knowledge and sometimes to the generation of new challenging questions.

6. *Dynamic involvement and animation*

These elements have been discussed at large in previous chapters. Let us just recall that it is essential to take part in the conversations and debates of the community, and eventually to provide financial means and technological platforms, and foster a supportive context to collective expression, open discussions and regeneration through the connections with other communities. In this regard, the manager should always be aware of the four components of the middleground and make sure to activate them regularly and consistently.

7. *Benevolent maintenance*

Even when the manager is not looking for specific elements of knowledge or solutions for the innovation projects of the firm anymore, she/he can preserve the potential of contribution of the community. This would mean remaining actively involved by participating, raising further questions, engaging in conversations and debates with expert members and newcomers as well. This is how the manager could benefit from potentially new ideas emerging from the knowledge development work of the community. At a minimum, the manager should monitor the activities of the community, which quite often can play an active role in prospective exploration on the evolution of a field of knowledge.

VIII. Conclusion

Communities are potentially a powerful source of innovation for firms. They can play a significant role in the development of the resilience and sustainability of firms. Yet, creating the right connections between the communities and the firm is a subtle endeavour, sometimes counterintuitive. The firm must understand and respect the motives and purpose of the community: its passion for the exploration of a specific field, its interest in developing knowledge further, its key purpose and values, over and beyond short-sighted economic interests. The firm must explore a new attitude of giving without expecting direct returns. Through genuine participation and support, it must learn how to foster potentially valuable exploration while letting go hierarchical control. This ability to engage in open-ended transversal dialogue with passionate people will be a key element for thriving in the 21st Century.

© 2021 World Scientific Publishing Company
https://doi.org/10.1142/9789811234286_bmatter

Postface: A Closing (and Opening Note) on Communities and "Commons"

Laurent Simon and Patrick Cohendet

In this book, we collectively tried to make the argument of the key role that communities could play in fostering and accelerating innovation, and we discussed different forms of communities in multiple contexts and specific cases. Before closing this book, we would like to refocus our thoughts and shed light on a key element at the core of what all communities are about: "commons".

Commons are a complex topic. They have been the subject of heated debates in economics and sociology for almost two hundred years, and the main contribution behind a recent Nobel Prize in Economics (Ostrom, 2009). We would like to give here a simple definition of commons as a set of goods that collectively belongs to a group of people that can be exploited by each member, and yet that requires care, maintenance and sometimes enrichment in order to be sustainable and to remain a part of the collective richness.

"Commons" are at the very heart of what communities are and how they operate. Members of a community are gathering around the existence, the maintenance and the development of their "commons". They thrive on the strong belief that their contribution and care are needed to keep the existence of a set of riches that expand over and beyond the individual and private property, and that can be enjoyed by all members of the community, and eventually be enriched by the expansion of the community to newcomers. They want to avoid and discourage "free-rider" behaviours, where members would enjoy the fruits of the work of the community without contributing to it, or at least taking care of it. A historical example could be the one of a commonly owned forest near a village. Villagers can freely enjoy the benefits from the forest, wood, berries, mushrooms or the pleasure of a walk in the undergrowth, under the shadows of trees. This would only be sustainable under the conditions of respectful, balanced and fair harvesting practices on the one hand, and of maintenance of paths, control of the wildlife, cleaning of the dead wood on the other hand. This requires awareness, respect and collective care.

A first takeaway from the fieldwork and different cases in this book is that these attitudes and behaviours of respect for the commons are in general naturally fostered, promoted and implemented by community members, not necessarily by firms and managers. Community members understand that the end-value of the commons will be potentially more important for all when everyone would respect them and engage in fair use, respectful exploitation and eventually enrichment of the commons through feedback and participation. It may be less obvious for firms that tend to look at the commons as another pool of resources. From this perspective, the actual challenge for firms, institutions and managers alike is to understand and respect the very nature of these sets of commons. It is to acknowledge their essence and to be as careful as possible when they want to enjoy some potential contributions offered by the exploitation of these commons. Commons shouldn't be approached as a set capital that one should acquire or appropriate, and own to exploit, excluding the others. They should be approached as fragile, organic sets of potential riches that should be exploited in fairness, and regularly maintained and sustained through participation, feedback on their use, new learnings, new elements of knowledge and eventually new challenges to regenerate the interest of the community. Firms must understand and acknowledge that fairly contributing and giving back to the community is not a generous act of benevolence, it is a necessary act of sustainability if they want to enjoy the contributions and fruits of the community in the long term.

A second takeaway is that commons may be more complex than we think. First introduced as shared pools of resources, then as shared knowledge bases, commons can be unfolded in multiple, interrelated dimensions.

What it means for a firm, an organization and managers who would like to be part of the life and experience of a community, is that, in order to learn form it, to generate new insights and to feed innovation processes, they should question and respectfully contribute to the following dimensions of commons:

— Common purpose
Communities are based and grounded in a shared understanding of the legitimacy of their exploration and quest for knowledge: a common purpose. As this purpose may be challenged by the development of knowledge or the evolution of the context of the community, it must remain open to debates, questioning and challenges. The firm must try to really understand this purpose, to engage in the "spaces" of the community, its meaningful questions, to check the alignment of its contribution with it and eventually contribute to its evolution through bringing its relevant knowledge into the discussion, and sometimes challenging it to foster debates, regeneration or pivoting.

— Places as commons
When we introduced the concept of middleground, we also emphasized that communities needed some "places" in order to thrive. With "spaces" defining the cognitive fields shared and explored by communities, the "places", physical or virtual locations, are the concrete, material and territorial playgrounds of the communities. The firm must identify the favourite place(s) of the communities and learn how to penetrate them without being too invasive. Here, firms can play an active and supportive role by providing these places — labs, living labs, workplaces, venues for events…, or virtual places, like forums or platforms. Again, supporting these places and helping in their organization and equipment, as well as the actual participation of the firm, are key elements in order to support the activities of the community.

— Knowledge commons
Knowledge is at the very heart of communities. Be it a field of expertise, a specific practice, the development of a theory or the exploration of a new style in art, members gather around a set of knowledge elements. They engage with it in order to feed it with their ideas or experience, share it and learn about it to consolidate it, translate it into activities and projects, and challenge it to keep it alive. In this regard, the way communities

thrive on knowledge illustrates the very nature of commons, as collectively owned, collectively maintained and collectively enriched, through social engagement, knowledge sharing and circulation, discussions and debates, exploration and experimentations. This challenges the firm to accept that it can't own knowledge with the traditional rights of control and appropriation, like patents or copyrights. The firm must find ways to feed the community back with learnings, experiences or new ideas to legitimize the possible exploitation of the work of the community.

— Social commons
Social connections are key elements for the emergence, unfolding and development of a community. Each new member not only brings some new elements of knowledge ("know about and know what") or inspiring practices ("know how") to the group, but also new possibilities of connections with potential new members and contributors ("know who"). This network effect is not automatic and requires handling connections in a very cautious fashion, as relationships are not mechanistic, but embedded in the context of shared personal stories, experiences and emotions. As the larger set of the connections of all the members can be seen as a powerful asset for the development, enrichment and legitimacy of the community, each introduction to a new connection remains personal and needs stewardship. That's why the firm must be represented by specific individuals, respectful of the personal histories and relationships. Also, the firm must also share with the group its specific valuable connections and make sure to support their introduction by respecting the rhythm and rituals of social relationships that are specific to the community.

— Values and behaviours as commons
As intangible as they may appear, the values fostered, practised and promoted by the community, and the behaviours that are driven and shaped by these values, can also be approached as "commons". Enacting these values through activities and the specific quality of relationships they channel is a core, distinctive element of the identity of the community. They can also be considered as commons, as their collective practice, maintenance and respect would ensure the cohesive role they would play and would consolidate the ties between members and their possibilities to interact, learn from each other and contribute to the collective endeavours. On the contrary, individual breaches in values and behaviours can appear as a threat to the whole group. Here again, the firm must cautiously learn

about values and the right behaviours. It should respect them as much as possible in order to be granted with legitimate participation in the community.

— Common projects
Projects are the activities through which the specific knowledge, learnings and ideas of the community are translated into action and put to the test of materialization. For members, engaging in a project is a way to demonstrate the value of the collective cognitive work, to challenge its relevance and evaluate its efficacy. It is also an essential way to complete the "know what" with the "know how". Participation in projects, as the enactment of collective knowledge, can also be addressed as a way to feed and maintain the community. Projects also appear as "commons", as the active participation of the more diverse set of members will help their concretization, acceleration and validation.

— Outputs and outcomes as commons
Finally, the fruits of the work of the community should be very obviously considered as "commons", at two levels:

(1) In terms of the outputs: the returns from the tangible results that can be harvested by any member, including the firm, should always be partially shared with the community. Reinvesting in the community is expected both as an ethical act and as a pragmatic one. Giving back is not only essential to respect the basic values of the community, it is a key concrete element to feed the "space", the development of knowledge, the evolution of core questions, challenging debates, and it contributes eventually to the material support of "places", "projects" and "events".
(2) In terms of the outcomes: if any exploration or project wouldn't necessarily generate economic and material returns, it would always feed the experiences of members. Sharing what has been learned throughout a project or an experimentation would always contribute to the evolution of the practices, processes or even orientation of the community. These meta-learnings, lessons learned, and takeaways can definitely contribute to the evolution of a community.

As the firm is generally the one that would exploit these outputs and outcomes, it bears a significant responsibility in managing this feedback toward the community, in order to ensure its sustainability.

It is only by acting as an honest and fair member of a community and by playing an active role — with humility, caution and somehow generosity — in the fostering, orchestration and maintenance of these multiple commons — that firms can hope for a status of legitimate participation in the community and eventually to get some significant and sustainable inputs for their innovation endeavours.

© 2021 World Scientific Publishing Company
https://doi.org/10.1142/9789811234286_bmatter

Communities of Practice: They Live by the Value They Produce!

An interview with Etienne Wenger-Trayner

by Karine Goglio-Primard

(Revue française de gestion, 46(287), mars 2020, p. 161–169).

Étienne Wenger-Trayner is a social learning theorist and consultant, a world opinion leader in the field of communities of practice and social learning systems. In his various works, he affirms human knowledge is a social act. He analyzes the links between community, knowledge, learning and identity of community members. He presents a theory of learning based on practice. By highlighting the role of communities, his learning theory helps organizations in all sectors face a number of challenges, such as designing more effective knowledge-based organizations, creating learning systems in organizations, etc.

Karine Goglio-Primard (KGP) — *You were at the origin of the concept of community of practice in your first book on situated learning, can you come back to your definition of a community of practice?*

Etienne Wenger-Trayner (EWT) — A community of practice is a technical term for a form of collective learning in which practitioners themselves discuss their practice. They reflect on what they do, make improvements, discuss problems, and help each other.

It is a community very much defined by the need to learn and the need to learn with peers — not just with researchers but also with peers who talk about their experience and use their experience as a curriculum. It is also their experience that allows them to understand a problem and try to address it together. There is an element of practice, an element of engagement in action that is very important to a community.

KGP: To continue on the properties of a community of practice, how is a community of practice structured?

EWT: There are a wide variety of communities. There are communities of practice that have very little structure. People meet on Fridays after work over a beer and talk. They don't say, we're a community of practice, they say, we talk. But if you listen to what is going on, you can see that there are a lot of discussions, problems to solve and improvements to be made in practice. It's very unstructured and, in some cases, it works very well.

At the other extreme, there are communities that are facilitated and sometimes even by professional facilitators. There is an agenda, a list of questions to discuss. It's very structured. There is often a manager who acts as a sponsor and who links the community directly to the company hierarchy.

There is a wide range of structuring of communities of practice that allows learning to take place between practitioners and in relation to the company. If people meet informally for lunch, the relationship with the company is through the community members who have projects. They bring the learning from the community into their projects without making a formal report. There are other communities that have a more formal process for articulating practices. They make a database, a kind of community memory. The degree of structuring depends on the company you are in and the degree to which communities are recognized by the company as a key element in its development and continuous learning.

KGP: Can certain communities be "piloted" by the company?

EWT: I have never understood the word "piloted", which comes from the French language. What is important in a community is that the members have a sense that it is their community. To the extent that what we call "pilotage" is something that takes away the sense of self-management in the community, for me, there is a danger that the community is just another meeting. One more meeting where we don't make the agenda. One more thing to do where we submit to the demands of the organization.

While a community that works well is different: there is a creative spirit. One person told me the community is the only place where I can "let my hair down". It's the place where I can be myself, where I don't have to do something that someone else asks me to do. It is an important part of who professionals are, that the community is their community. Now the community serves the business. The company can ask practitioners questions: we have this problem, what do you think? This confirms the sense that the community is a place where I am myself and where my intelligence is recognized as a contribution to the company. The community needs to be a place where you can engage your intelligence and perspective. A place of agency: a place where the perspective of practitioners is valued, where it has an effect on what happens in the company.

Often employees in companies do things that don't necessarily reflect their perspective but reflect the perspective of a business process or of a manager. If what you call "piloting" is controlling, then it's more like a focus group. In a focus group, we are very interested in the idea of people. But the questions are asked by the researcher and the final conclusions are the researcher's business. A focus group asks people for their perspective, but the control is not with the people. I'm not saying that this self-management is limitless in a community of practice. Practitioners from a community in a company are employees of that company. They use the time for which they are paid by the company to participate in the community. What they do together must be of benefit to the company.

I've never seen a community where what's happening is completely at odds with the company. But I'm sure it exists. If there is a lot of mistrust in a company, the community can become a rebellious place. But if communities of practice become rebellious places, I think it's really very important for the company to ask why. Why do they need to be rebellious? There has to be a design problem in the company for employees to need to be rebellious.

KGP: Can the company put forward questions to a community?
EWT: Yes absolutely. I studied a community that was angry with the company. The company had made a poor acquisition of a small business in the community's area of expertise without consulting community practitioners. You have a community that can give you that perspective of practice, why not ask them for their opinion. Community members didn't ask to make the decision. They didn't want to have the power of decision because the power of decision would probably make their community too

political. It is better that the final decisions are made in the hierarchy but that the community is consulted. For them it was just common sense.

KGP: If the community is consulted by the company, it can motivate the members. What are the other keys to motivating a community?
EWT: A big key to motivating a community is that the discussions focus on the real problems of the practice. This is what this person meant when she said I can "let my hair down". Now I can really talk about real problems and in a real way. That's why some communities have decided that managers can't participate. We are often asked whether it is good for managers to participate in the community or not? It depends on the case.

For example, in consulting firms, it is often good for managers to participate because they tend to be practitioners themselves and their opinion is important. If the managers do not participate, the conclusions reached will not be taken into account. While participating in the community, managers are not really managers. They are practitioners with the people in the community.

But in another company, for example, the communities had decided that managers would not participate in engineering communities. The engineers said that if their manager was there, then there are all kinds of things about their practice that they couldn't say. In fact, the managers formed their own community.

In a community, you have to be with people who understand why what you're saying is a problem. Why a question you ask, an uncertainty you have is a real thing and a difficult thing. Over time, they may invite managers for some discussion. A community may have people who participate by invitation but not as full-time members. There is no general recipe. You have to find out what works well in each company to create this link between communities and the company.

KGP: What is the facilitator's role in motivating community members?
EWT: The best thing is for facilitators to be practitioners themselves. They don't need to be experts in the domain. But they need to be practitioners who understand the practice, have questions themselves, and are able to focus on the key issues to address for members to find value in participating. In a sense a community survives on the value it produces.

The facilitator's main role is also to ensure that for those who participate it is always a good investment of their time and that the value for time equation is in favour of value. The community really needs to be focused on what produces value for the members because value for the members is also value for the business. In this sense, facilitators provide a strong link with the company.

KGP: Yes, the facilitator is often a manager or director, is this the link with the formal structure?
EWT: But I don't think it is necessary for facilitators to be at a certain level in the hierarchy. They have to be respected practitioners …

KGP: Yes, legitimate practitioners.
EWT: The issue of legitimacy is important. And if a manager has that, then it's fine, but often being a manager is not enough legitimacy to be the engine of a community.

KGP: What is the life cycle of a community?
EWT: There is so much variation. There are communities that live for years, decades, even if they have ups and downs. Others have a much shorter duration.

There is often a degree of tentativeness at the beginning of a community. Will this give me value? There is a bit of that sense of exploring the relationship. Is there good chemistry between us? Do I find it a good investment of my time? Once this is established, the community can set rules or ambitions. But I would say it is something that should follow practice rather than precede it. It's very different from a team. You form a team around a well-defined work plan in advance, and the value is at the end of the project. The problem with communities is that if you come in with a big work plan, people say I don't have time for that. You have to be careful not to scare people in advance.

As I said, a community is a structure that lives by the value it produces. Often in a company, community participation is voluntary. (Not always, but usually.) If participation does not produce enough value, in the end, people simply become less invested in the community.

KGP: Is the community a new form of management for business organizations? A non-hierarchical, more flexible, cooperative, almost semi-formal structure?
EWT: Community is a structure that helps to fill in the gaps in corporate structuring. Belonging to a community does not depend on which department you are in or who your boss is. It depends on your relationship with the practice. So, it allows a more flexible structure.

A company structure is usually organized on one axis, sometimes two, but it is never enough. For example, even if the company is organized by business line (by products, by country, by function…), there is always a main axis and sometimes a secondary axis. But these matrices are not capable of capturing all the forms of relationships that are important to the functioning of the company. Communities fill in the interstices.

If you're organized by product, for example, you could have all the people involved in the manufacturing process forming a community. By crossing product lines, this allows you to create relationships that are recognized by the company but are different from the main axis.

If you are organized by country, it is important that people who do the same thing in different countries have a relationship that allows them to learn from each other, for instance, to avoid repeating the mistakes that have been made elsewhere in the company.

KGP: This allows us to identify new and more transversal forms of organization. Do communities break down organizational silos?
EWT: Silos have a reason to exist. It's that humans only have one brain and 24 hours a day. A company with communities of practice is not necessarily a company where there are no silos. For example, when managing budgets, it is important to have silos where responsibilities are clear. It is important to have teams to take care of projects and formal structures to manage budgets. A community of practice is not a well-suited structure to manage a large budget. There was a movement in the business to flatten organizations — to have fewer layers of management between the workers and the CEO. But you don't need to have a flat organization to have communities of practice that work well, because the communities are transversal anyway. In some military organizations that are very hierarchical, there are communities that function well, precisely because they are structured transversally. This does not mean that there is a need for a very vertical and very high hierarchy. It simply means that the viability of the communities does not depend on one or the other.

In my experience, communities work well when there is a commitment on the part of the company to recognize them in a sustained way.

One of the problems I've seen with communities in formal organizations, whether it's in business, in government or in international development, is a tendency to promote communities for a few years … and then nothing. And then it comes back. The interest in communities on the part of management comes in waves, and this has a very negative effect on the development of communities.

The communities don't ask for much. A small budget to travel, to meet face to face once a year. An executive sponsor who takes the community under his or her wing. These are good things, but it has to be long term. If it comes and goes...it's not good because people get involved in a community, their identity gets emotionally invested. People need enough recognition so that the investment they make in their community is recognized by the company as having value.

Organizationally, it's good that the community has someone to talk to, someone in the hierarchy who understands what they are doing and the value it has for the company. The simplest mechanism is having a sponsor who provides this very important link between the community and the company. If the community has a very good idea, they need to know where they can present their idea. Many ideas coming out of the communities are things that practitioners themselves can implement in their own work without going through the company; but sometimes more consequential ideas need the support of the hierarchy to be realized.

KGP: The community can be solicited by the company to solve problems related to the company's activity, examples?
EWT: Yes, but it has to be from the practitioners' perspective. A community of practice is not a taskforce. For example, if you want to change the IT system in a company, then you set up a taskforce. Representatives from different parts of the company study the problem and recommend a solution. It's not a community of practice because you need different perspectives for this taskforce to come to a good conclusion that is accepted by everyone.

But for example, I have seen a community that took on an efficiency problem in a metallurgical company. In agreement with its sponsor, the community had decided to explore what people in the world are doing in their field. Their idea was to find a way to reduce the energy consumption of certain systems. It was very interesting to see how the community was inspired by this collective exploration and was proud of the results.

Since communities in general are voluntary, the question is not whether they are given a task or not. The real question is: is the task going to inspire the members or is it going to be one more thing to do?

KGP: How do communities contribute to innovation in organizations?
EWT: I've never been asked in my work to integrate communities with the formal business innovation process, so I can't address that aspect. First of all, we should not romanticize communities. Communities can be a place of resistance to innovation, especially if people feel that innovation is a threat to their practice. But communities are sources of innovation in all sorts of ways. It's important that people have a place where they can freely discuss an idea, if it is half-based, explore it, talk about it with others and get their perspective. Because often in a team, if you have a very new idea, the team has to finish its project and there is no time to explore speculative ideas. The community is a place of discussion where you don't have to finish something very quickly, as is often the case in a project. It is an informal place where there is no pressure to finish a project for the client. You can explore things where you are not sure of the result and this is very important for innovation.

KGP: Yes, the community is a good place for creativity, idea development and innovation.
EWT: Sometimes communities can be used for brainstorming to solve an urgent problem. A member will say I don't know how to respond to my client's request. Is this a problem you have faced before? But also, a community can be a place where people say: I've thought about this, what do you think about it? Is it worth thinking about it more?

KGP: Often the company has difficulty taking advantage of this community creativity. Ideas will stay at the community level.
EWT: Are these companies where there is a sponsor?

KGP: No, often these are companies where there is no sponsor.
EWT: One of the important roles of the sponsor is to listen and make sure that good ideas don't get stuck in the community. In the example above, the engineers didn't want the managers to be involved in their community. So, managers made their own communities to help each other as managers and also to become sponsors of 5 to 10 engineering communities. There were often conversations between these two types of communities.

KGP: In our observation of communities of practice, we found several types of communities: communities of experts, communities of knowers

and unknowers, and communities of problems. From your point of view, are there several typologies of communities of practice?
EWT: Absolutely. In fact, communities that mix the knowledgeable and the unskilled are an interesting example. I've often heard that these are communities where the knowers are gradually disappearing. If they have little time, the knowledgeable experts want to spend that time dealing with difficult problems rather than repeating things for beginners.

KGP: These communities of the knowers and the unknowers are on the strategic axis of the company. That's why it works well.
EWT: Yes, probably communities, I guess, where experts benefit from the people they interact with having a good knowledge of this strategic area.

KGP: Yes, the people they interact with have customers that are important to the whole company. So, it's a way to get the strategic area known to customers around the world.
EWT: There are also communities of experts that are closed, by invitation only. For example, in a large consumer products company, there were communities where you became a member by invitation only. It was an honour, and it worked well.

There are open expert communities where non-experts can come, but they shouldn't interrupt the discussions with their beginner's questions. They come to better understand how experts are addressing issues of interest to them.

There are also communities where experts take care of the knowledge of non-experts. There are even communities where experts are paid by the company to create learning opportunities for non-experts within the community. The idea is that this is not very motivating. So, there is a budget for these experts to spend time helping newcomers.

KGP: What are the major pitfalls in the functioning of a community?
EWT: One of the major pitfalls is time being neither recognized nor valued. Without recognition and legitimacy, it becomes time that members have to steal from their work rather than time spent as an integral part of it. And there are more specific pitfalls, such as giving tasks that are not motivating to a community. There are also interpersonal conflicts with someone who is always attacking people. There are all kinds of pitfalls because it's a piece of a community. A community is a delicate construct.

KGP: What good practices have you observed or experienced in a community?
EWT: I don't know if it's a practice but it's the idea that the community is self-managing, that there are ways for people to gain a sense that this is their community. For example, having a little bit of time to invite members to think about how the community can create value, processes that allow for a real sense of self-management.

Another good practice is the so-called "case clinic": a member comes with a problem, and the community uses this problem to learn together. Pushed by a real problem from one of the members, everyone learns. It is not just the person who brings the problem to the community who learns. Everyone learns by seeing how others react to the issue. There is interaction not only with the person bringing the problem but also among all members of the community. And often in organizations, a good practice is to have someone who has some time dedicated to cultivating the community.

KGP: Is this what we call in French the "animator"?
EWT: Yes, you could call it that. But the word "animator" makes it sound like the community is dead and needs to be animated. It just needs someone who can spend some time making sure that the community is working well and that it is producing value for the members.

KGP: A participant that plays a facilitator role?
EWT: Yes, someone who has some time set aside to maintain the community space. Community members have just enough time to participate in the community, but not enough time to worry about its ongoing functioning.

This is often also the person who will have a relationship with the sponsor. It should be pointed out that there are several types of sponsors:

- There can be a sponsor from the communities in general in the whole organization.
- There is also the sponsor of a specific community: it is often a manager whose line of business depends on the quality of what the community does.
- And sometimes it takes the individual sponsor of a member. A good practice for a facilitator is to send an email to a member's manager to explain what that member has done for the community. It is not in the

formal description of their job, but it is an important contribution. So, the manager will realize that the community is not a waste of time, but it is really contributing to the company. The community will also benefit that manager's business unit.

KGP: What are your activities in California and Portugal?
EWT: We've started a Social Learning Lab in Portugal, where we get people from different organizations in various sectors to meet and learn together. We don't organize educational seminars. We convene all kinds of mutual learning encounters. The idea of these meetings is, over time, to develop the practice of social learning as a way to address more effectively the important challenges of today's world for individuals, organizations and society.

© 2021 World Scientific Publishing Company
https://doi.org/10.1142/9789811234286_bmatter

Index

K

L

M